T0228166

Flux Bounded Tungsten Inert Gas Welding Process

Flux Bounded Tungsten Inert Gas Welding Process

An Introduction

By

P. Chakravarthy, M. Agilan and N. Neethu

CRC Press
Taylor & Francis Group
Boca Raton London New York

CRC Press is an imprint of the
Taylor & Francis Group, an **informa** business

CRC Press
Taylor & Francis Group
6000 Broken Sound Parkway NW, Suite 300
Boca Raton, FL 33487-2742

Library of Congress Cataloging-in-Publication Data

Names: P., Chakravarthy, author. | M., Agilan, author. | N., Neethu, author.
Title: Flux bounded tungsten inert gas welding process : an introduction / Chakravarthy P, Agilan M, and Neethu N.
Description: First edition. | Boca Raton, FL : CRC Press/Taylor & Francis Group, 2020. | Includes bibliographical references and index.
Identifiers: LCCN 2019042487 (print) | LCCN 2019042488 (ebook) | ISBN 9780367422875 (hardback) | ISBN 9780367823207 (ebook)
Subjects: LCSH: Gas tungsten arc welding.
Classification: LCC TK4660 .P234 2020 (print) | LCC TK4660 (ebook) | DDC 671.5/22--dc23
LC record available at https://lccn.loc.gov/2019042487
LC ebook record available at https://lccn.loc.gov/2019042488

Visit the Taylor & Francis Web site at
http://www.taylorandfrancis.com

and the CRC Press Web site at
http://www.crcpress.com

Contents

List of Figures

List of Tables

Preface

WELDING AS A TECHNIQUE in manufacturing for joining distinct components has been in practice for a long time. Ever since it has been known to mankind, there have been many technological advancements to suit its applications. Though tungsten inert gas welding is very successfully practised in industries, many variations have been introduced by the scientific community. This monograph intends to briefly explore the fundamentals of tungsten inert gas welding processes and the variations that have been introduced in the recent past, depicting the applicability of various processes in industries and the mechanisms involved, with particular focus on flux bounded tungsten inert gas welding. It is designed primarily for practising engineers in industries and graduate and postgraduate students in the field of metallurgical and materials engineering, mechanical engineering and production engineering, but the content is suitable for undergraduate students also in these disciplines.

This monograph is structured to introduce the concept in a coherent way through four chapters. Chapter 1 emphasizes the need for welding, highlights the chronological developments in tungsten inert gas welding and the features of a fusion welding process, while Chapter 2 throws light on the various aspects of tungsten inert gas welding, including its variations, equipment, process parameters, etc. The concept of activated tungsten inert gas welding is presented in Chapter 3, and Chapter 4 is dedicated to flux bounded tungsten inert gas welding process. The process parameters, the applicability of the process in industries, pros and cons of this process over the other equivalents and the mechanisms involved are discussed in detail.

There has been a strong motivation behind writing this book, which was kick-started by initial discussions with Shri A. V. Santhana Babu (scientist/engineer at the Indian Space Research Organisation), who demonstrated the viability of flux bounded inert gas welding process for welding aluminium-based components. The authors are very grateful to Shri S. Jayakrishnan (scientist/engineer at the Indian Space Research Organisation) for conducting the initial studies on this technique, which provided a platform to explore it further. The authors also thank Shri D. I. Arun (scientist/engineer at Indian Space Research Organisation) for the deliberations and his help in shaping this monograph.

The lead author, P. Chakravarthy, would like to express his sincere thanks to his revered parents, Shri Padmanabudu and late Smt. Rajini Kumari, for all their sacrifices. He also conveys his special gratitude to his spouse I. Naveena for her utmost forbearance, without which this monograph would not have been completed. The support from his friends Shri T. V. Joseph and Shri Ezhilan is greatly acknowledged.

The author M. Agilan would like to express his sincere gratitude to his mother, Chinna Penn; wife, Smt. Poorni; his son, Adhitya; and his brother, Shri Ezhilan for their continuous encouragement and time.

The author N. Neethu would like to express her gratitude to her parents Mr. K. M. Nazar and Mrs. C. R. Nadhi for their constant support and her elder brother, the late N. Nidhin, for being her guiding light and inspiration. She would also like to express her sincere thanks to every teacher who has taught her, for all their contributions and sacrifices. She also acknowledges the help and support of her friend K. Nahil Ahmed Hassan.

P. Chakravarthy
M. Agilan
N. Neethu

Authors

P. Chakravarthy is a faculty member in the Department of Aerospace Engineering at the Indian Institute of Space Science and Technology, Thiruvananthapuram, India. He holds a doctorate in metallurgical and materials engineering and has vast experience in the field of material processing. His current focus includes the development of newer materials and processes for the aerospace sectors.

M. Agilan is currently working as a scientist in the Vikram Sarabhai Space Centre, Indian Space Research Organisation (ISRO), Thiruvananthapuram, India. He has nine years of experience in welding and joining of aerospace materials. He has been working on gas tungsten arc welding, electron beam welding and friction stir welding of various strategic materials. He has completed a BE in metallurgical engineering from the Government College of Engineering, Salem, and is currently pursuing his doctoral research work in the Indian Institute of Technology, Chennai, India.

N. Neethu is a final year undergraduate student in aerospace engineering at the Indian Institute of Space Science and Technology, Thiruvananthapuram, India. Her primary research interest lies in the fields of materials science, metallurgy and manufacturing. She has expertise and experience in welding and forming and has publications in these fields.

Nomenclature

γ	Surface tension per unit length
AATIG	Advanced activated tungsten inert gas
ATIG	Activated tungsten inert gas
C	Concentration of surface-active element
DCEN	Direct current electrode negative
DCEP	Direct current electrode positive
DCRP	Direct current reverse polarity
DCSP	Direct current straight polarity
DWR	Depth-to-width ratio
ESW	Electroslag welding
FBTIG	Flux bounded tungsten inert gas
FCAW	Flux cored arc welding
FZTIG	Flux zoned tungsten inert gas
GMAW	Gas metal arc welding
GTAW	Gas tungsten arc welding
GTCAW	Gas tungsten constricted arc welding
HAZ	Heat-affected zone
PAW	Plasma arc welding
P-GTAW	Pulsed gas tungsten arc welding
PMZ	Partially melted zone
S	Distance
SAE	Surface-active elements
SAW	Submerged arc welding
SMAW	Shielded metal arc welding
T	Temperature
TIG	Tungsten inert gas

CHAPTER 1

Introduction

WELDING HAS BEEN RECOGNIZED all over the world today as one of the most versatile means of fabrication processes. It has a significant role in almost all manufacturing industries due to the inability of conventional manufacturing techniques to fabricate large components or products as integral units. Instead, most industries machine out smaller parts of each product and then depend on different procedures to merge or unite these parts. Welding addresses the difficulty in manufacturing and transporting large components as single integral units as the process helps to join individual parts to the final integral product. Generally, it is a fabrication process that joins materials by coalescence due to heat. In most cases, since the component's behaviour and its effectiveness depend on the joint strength, temporary joints are generally avoided due to the innate restriction in providing sufficient strength to the joint. Among all other manufacturing processes (casting, machining, forming and powder metallurgy), welding and its allied processes are essential for the manufacture of a range of engineering components to produce a complex assembly.

Ever since it was developed, welding or joining has been widely used to make tools and structures. Welding, by its classical definition, produces coalescence (joining) of materials by heating them to suitable temperatures with or without the application of pressure or by the application of pressure alone, and with or without the use of filler material. Thus, welding ensures continuity between parts for assembly, by various means to transmit load/power and in some cases to restrict the degrees of freedom of components. Continuity in this context not only implies the homogeneity of chemical composition but also the continuation in the atomic structure. After the parts are joined by a welding process, the two separate entities that existed before become one single entity. In welding, a filler material is sometimes needed to facilitate coalescence.

The first evidence of welding can be traced back to the Middle Ages. Broadly, welding can be defined as a process to achieve a metallic bond. In this sense, examples of products obtained by welding can be traced as far back as the Bronze Age, around 3,000 BC. Archaeologists have unearthed jewellery made by hard soldering and swords produced by hammering. However, much to our surprise, the growth of welding to become accepted as a conventional industrial process took nearly 5,000 years. It is widely believed that the art of welding started with iron, when the process of smelting ores of metals became popular. During the Iron Age, forge welding came into its own. The technique of joining metal parts by heating it to a dull red colour and hammering (pressing) them together was practised as a method of joining in the primitive ages. This became the traditional hammer forging process of the village blacksmith, which in common terms today is a forge welding process. The age-old iron pillar at New Delhi, India, made by this process of pressing the iron ingots together exemplifies the work of the skilled personnel performing such processes at that time. It is also noteworthy that the well-known Damascus swords and many other specimens of ancient and medieval swords that are seen in various museums were produced by the technique of forge welding. Since then, there have been many developments in joining of materials based on the requirements. As this book emphasizes flux bounded tungsten inert gas (FBTIG) welding, a modified tungsten inert gas (TIG) welding process, the chronological developments in the TIG welding process are mentioned in the following section.

1.1 CHRONOLOGICAL DEVELOPMENTS IN TIG WELDING

- In 1800, an arc between two carbon electrodes using a battery was produced by Sir Humphry Davy.

- During the middle of the 19th century, the electric generator was developed.

- In 1890, C. L. Coffin patented gas tungsten arc welding (GTAW) in a non-oxidizing gas atmosphere.

- In 1920, P.O. Nobel of the General Electric Company invented automatic welding.

- In 1926, H. M. Hobart used helium and P. K. Dever used argon as a shielding gas in TIG welding and received patents independently.

- In 1941, Meredith (USA) perfected the process and named it Heli-arc welding.

- In 1957, Gage invented the plasma arc welding (PAW) process where a constricted arc plasma is produced for welding and observed that the arc temperature in PAW is much higher than the tungsten arc.

- In 1960, the Paton Welding Institute (PWI), Ukraine, first reported that the use of activating fluxes improves the performance of the TIG welding process.

- In the 1970s, a transistor-controlled inverter welding power source was introduced.

- In 1964, the 'hot wire' welding process was developed and patented by Manz.

- In 1965, predominantly TIG-welded components were used in the Apollo 10 spacecraft.

- In the 1980s, semiconductor circuits and computer circuits used to control welding and cutting processes were developed.

- In the 1990s, inverter technology dominated power supply designs and led to reduced size and weight of welding power sources.

- During the last five decades, several advancements and modifications in TIG welding have been made, specifically in the area of power sources, automation and defect control to improve process efficiency, safety, etc.

1.2 CLASSIFICATION OF WELDING PROCESSES

Welding is classified based on the physical state of metal and the metal flow during welding. It is classified as

- Fusion welding.

- Solid state welding.

According to the topic chosen, this discussion is confined to the arc welding process, which is categorized as one of the fusion welding processes.

1.3 FUSION WELDING PROCESSES

In these processes, the faying surfaces of the parent metal and the filler metal (if required) melt and form an integral joint, which involves the fusion of the edges of the base metals to complete the weld. Fusion welds ordinarily do not require the application of pressure, and they may be completed with or without the requirement of filler metal. The requirement of filler metal generally becomes a necessity only when the thickness of the base metals to be joined is large enough, usually greater than 3 mm. Usually, fusion welding processes use a filler material to ensure that the joint is filled. The heat for fusion is supplied by various methods, and one of the common methods is through electrical energy. In arc welding, an alternating current (AC) or direct current (DC) power unit capable of supplying a high current but low voltage is used, and the arc is struck between an electrode and the base metal. Arc welding covers most of the welding processes under the fusion welding category. Though electron beam welding and laser beam welding are fusion welding processes, they are not categorized as arc welding processes because of the nature of the heat source.

The welding arc is a high-current and low-voltage electrical discharge which flows from the cathode to the anode. The flow of current through the gap between the electrode and the workpiece needs a column of charged particles to have reasonably good electrical conductivity. The electric discharge is sustained through a path of ionized gaseous particles called plasma. Various mechanisms such as field emission, thermal emission, secondary emission etc. cause the generation of these particles. The temperature inside the arc and at the surface of the arc is approximately 15,000°C and 10,000°C, respectively. The open-circuit voltage for a typical arc welding process ranges from 30 to 80 volts, and typical currents are between 50 and 300 A. The energy developed in the arc per unit time equals $V \times I$, where V is the arc voltage and I the current. The welding arc acquires the shape of hot gas formed between the electrodes, and due to its low density, hot gas tends to rise and form a bell-shaped arc. Further, fusion welding processes are categorized based on the type of electrode used.

- Consumable electrode processes – shielded metal arc welding (SMAW), submerged arc welding (SAW), flux cored arc welding (FCAW), gas metal arc welding (GMAW) and electroslag welding (ESW) processes.

- Non-consumable electrode processes – gas tungsten arc welding and plasma arc welding processes.

A brief outline of the commonly used fusion welding processes is given below.

1.3.1 Shielded Metal Arc Welding (SMAW)

Among arc welding processes, SMAW is the most common, economical and versatile process for the fabrication of structures throughout the world due to its equipment portability and the availability of a large range of consumables. This process is also known as manual metal arc welding or stick welding in layman's language. In SMAW, a consumable electrode usually coated with a suitable flux is used to generate the electric arc. Due to the intense heat of the arc during welding, the electrode melts to form droplets and is transferred to the base metal. The decomposed gases from the flux shield the arc, thereby protecting the molten droplets, and the base metal is protected by the slag cover. The coating material performs several functions:

a) Protects molten metal from oxygen and nitrogen.

b) Helps to stabilize and maintain the arc.

c) Aids in deoxidation and weld metal refinement.

d) Modifies the composition of the weld metal by alloying addition; thereby desired mechanical properties and microstructure are achieved.

e) Controls weld bead profile and weld spatter.

f) Increases weld penetration.

Major constituents in the flux coating are cellulose, sodium and potassium silicates, metal carbonates, rutile, ferromanganese, ferrosilicon, limestone, etc. Depending on the coating design, SMAW can be operated with either positive or negative electrodes using a DC or AC power source. Arc initiation in SMAW is achieved by electrode the 'touch start' method or 'drag' method, and after arc initiation, a proper arc length is be maintained to achieve a good weld.

SMAW can be used for almost all common metals and alloys. It is employed for fabrication, assembly, maintenance and repair work and field construction due to its simplicity and portable and inexpensive equipment (power supply, electrode holder and cables). The process has some significant disadvantages as well.

Compared to welding techniques that uses inert gases to protect the arc, the shielding protection is inadequate here. Also, the deposition rates are lesser compared to other arc welding processes. It also requires skilled operators since it is mostly performed manually, rather than automatically.

1.3.2 Submerged Arc Welding (SAW)

In the submerged arc welding (SAW) process, the arc is established beneath a hill of granular flux particles, which makes the arc invisible. The flux protects the arc and the molten weld metal from the ambient atmosphere, thereby preventing the formation of oxides and other adverse reactions. The flux is supplied from a hopper which travels along with the torch. Since the molten metal is separated from the atmosphere by the molten slag and granular flux, no shielding gas is required. Sometimes, base metal powders and alloying elements are added to the granular flux to increase the rate of deposition and to control the composition of the weld metal. Due to the high deposition rate in SAW, high-thickness welds can be produced.

1.3.3 Tungsten Inert Gas (TIG) welding

Tungsten inert gas (TIG) welding is also known as gas tungsten arc welding (GTAW) or Heli-arc process. In this process, an electric arc is maintained between a non-consumable tungsten electrode and the part to be welded. The molten metal is protected by a blanket of inert gas fed through the welding torch. Inert gases such as argon, helium or an argon–helium mixture are used. Though it was developed to weld aluminium and other reactive metals, now the process is used for almost all metals and their alloys. Tungsten is used as the non-consumable electrode due to its high melting point, low electrical resistance, good heat conductivity and the ability to emit electrons. Additives such as thorium, zirconium and cerium are added in the range of 1–2% to improve the performance of the electrode. Because of low heat input, thin sections are easily welded. As a clean welding process, reactive metals such as titanium, zirconium, aluminium and magnesium are capable of being welded. TIG welding is used in critical applications such as in the aerospace, defence and nuclear industries, etc., due to the ability to produce high-quality welds. However, the deposition rate of this process is low.

1.3.4 Gas Metal Arc Welding (GMAW)

In gas metal arc welding (GMAW), the arc is established between a consumable electrode and the workpiece. The electric arc and the molten metal are shielded by inert gases (Ar and He) and CO_2. GMAW uses the direct current electrode positive (DCEP) mode to obtain a stable arc, smooth metal transfer and good weld penetration. Constant arc voltage characteristic is normally employed in GMAW. In this process, molten metal from the electrode to the weld pool can be transferred by three basic transfer modes, namely, short circuiting, globular and spray transfer. Short circuiting transfer occurs with low current and while employing an electrode of smaller diameter. When the molten metal from the electrode tip touches the weld pool surface, short circuiting occurs. This mode is preferred for welding thin sections and out-of-position (overhead) welding. Globular transfer takes place when the metal drop size is greater than the electrode diameter, and under the influence of gravity, the droplet detaches from the electrode and transfers to the weld pool. Spray transfer occurs at a high current level, wherein electromagnetic forces cause the small discrete metal drops to move through the arc gap. Spray transfer mode is more stable and produces welds more spatter-free than the globular mode. GMAW is widely used for welding of aluminium alloys, and a high deposition rate can be achieved for welding thicker sections at a higher welding speed.

1.3.5 Plasma Arc Welding (PAW)

Plasma arc welding (PAW) is an advanced variant of TIG welding in which a constricted arc is established between the tungsten electrode and the workpiece. Arc constriction is achieved by passing the arc through a water-cooled orifice gas nozzle. Therefore, both shielding gas and orifice gas are used in this process. In PAW, the pilot arc is ignited between the electrode tip and orifice gas nozzle, and subsequently, the electric arc is transferred to the workpiece. Unlike melt-in-mode in the conventional arc welding process, keyhole mode is also achieved in PAW. Due to this advantage, high-thickness sections can be welded in a single pass.

1.4 VARIOUS ZONES IN FUSION WELDS

A fusion weld joint is characterized by a fusion zone, a weld interface, a heat-affected zone and an unaffected base metal zone.

Fusion zone: This is characterized by the as-cast solidified microstructure. The adjoining plates act as a metallic mould wherein the molten metal solidifies and the applied heat is extracted by them. As the weld region is small, the rate at which the heat is extracted by the adjacent base metal is large, thereby resulting in higher solidification rates, which are usually greater than those of castings. Therefore, fine grain microstructure is achieved. The final chemical composition of the fusion zone depends on the type of filler metal used and its dilution.

Weld interface: This is a narrow boundary between the weld fusion zone and the heat-affected zone. Generally, in the fusion zone, dendrites nucleate from the interface and grow towards the weld centre. Sometimes the interface has a partially melted zone (PMZ) whose formation depends on the weld metal composition. Alloys with low melting phases like eutectics have a PMZ more often.

Heat-affected zone (HAZ): This is the region of the base metal which is heated below its solidus temperature, and microstructural changes occur due to the heat of welding. The width of the HAZ depends on the heat source, applied heat input, the thermal conductivity of the base metal, etc. The microstructure of HAZ is affected by the heating rate, cooling rate and the peak temperature influenced.

Unaffected base metal zone: This is the region of the base metal next to the HAZ, which is not affected by the heat of welding. But the region much nearer to the HAZ will have high residual stresses.

CHAPTER 2

Tungsten Inert Gas Welding

THE TUNGSTEN INERT GAS (TIG) welding process is one of the most widely adopted processes in industries for joining similar/dissimilar metals. The increasing demand for weld quality has made this process very popular for welding smaller sections to huge pressure vessels for critical applications. TIG welding has become an essential process in industries because of its versatility and its ability to produce high-quality weld joints with low equipment costs. In the early 1920s, the possibility of shielding the welding arc and the weld pool using helium was first examined. However, significant progress was not made until the beginning of World War II, when there was a huge demand in the aircraft industry to replace riveting for joining reactive materials such as aluminium and magnesium. During World War II, this process was developed to weld reactive metals; later it was extended to weld a wide range of materials like stainless steels, nickel alloys, etc. Using a tungsten electrode and direct current (DC) power source, a stable and efficient heat source (arc) was produced, which resulted in excellent welds. Helium gas was selected to provide the

11

necessary shielding since it was the only readily available inert gas at that time. This process is also known as the Heli-arc process or non-consumable electrode welding process. The terminology given by the American Welding Society (AWS) is gas tungsten arc welding (GTAW) because other shielding gases that are not inert are also used in some applications.

This process uses the heat generated by an electric arc between the non-consumable tungsten electrode and workpiece to melt the faying surfaces. An electric arc is established between the tungsten electrode and workpiece by passing an electric current through a conductive ionized shielding gas. A schematic representation of TIG welding is shown in Figure 2.1. The molten metal, heat-affected zone and the tungsten electrode are protected from the atmosphere by a blanket of shielding gas fed through the TIG torch. Also, the shielding gas provides the required arc characteristics. Based on the thickness of metal to be joined, the filler metal may be added or autogenously welded.

Since the development of TIG welding, several improvements have been made to the process and its equipment. Certain welding

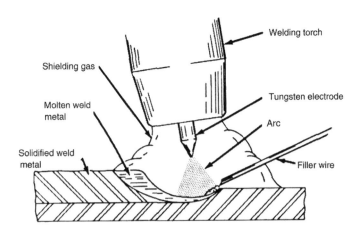

FIGURE 2.1   Schematic representation of tungsten inert gas (TIG) welding (reprinted from Antonini, J.M., *Comprehensive materials processing* (2014), Vol. 8, with permission from Elsevier).

power sources have been developed specifically for this process which includes pulsed DC and variable-polarity AC power source. Water-cooled and gas-cooled torches were also developed. The tungsten electrode is alloyed with small amounts of active elements to improve its emissivity, arc initiation, arc stability and electrode life. Various compositions of shielding gas mixtures were investigated and found to have better performance. Areas such as automatic controls, current pulsing, arc length controls etc. are still being actively studied.

2.1 WELDING POWER SOURCES

Arc welding power sources are categorized based on their output characteristics with respect to voltage and amperage. They can be either constant current (CC) or constant voltage (CV) or a combination of both. Constant-current power sources are used in processes such as TIG, SMAW and SAW where constant current flow in the welding circuit is maintained irrespective of voltage variation (arc length). These processes require the welder to maintain the arc length rather than the equipment. In CC power sources, a relatively larger change in arc voltage produces a small change in output current. They are also known as drooping power sources since they produce substantial downward slope (negative slope) of the voltage-amperage curve. Constant-current power sources can be inverters, transformer-rectifiers or generators. The transformer-rectifier and inverter power sources transform input AC power to DC power as output whereas generators convert the mechanical energy into electrical power. Normally, TIG power sources are transformer-rectifier, constant-current, AC/DC type. In certain cases, inverter power sources are also used in TIG welding which uses an inverter circuit to produce a high-frequency AC, reduces that voltage with an AC transformer and rectifies that to obtain the required DC output.

Pulsed gas tungsten arc welding power sources are used to improve the efficiency of TIG welding. The objective of pulsing is to repeatedly heat and cool the weld pool. The peak current depends on achieving a proper weld pool size during the peak pulse time without melt-through. The background current and duration are based on achieving the required rate of cooling of the weld pool. In the background-current cycle, the solidification rate is accelerated and weld pool size is decreased without disrupting the arc. Consequently, pulsing results in increasing and decreasing the size of the weld pool alternately. The open-circuit voltages (OCVs) of constant-current rectifier power sources vary from 50 V to 100 V based on the welding application. GMAW and flux cored arc welding (FCAW) processes use the constant voltage power source where the voltage is maintained at the set value irrespective of the current used in the process.

2.1.1 Duty Cycle

During welding, the continuous flow of current through various components of power source leads to overheating of transformer winding and other components leading to reduced performance. The duty cycle is a ratio of the load on time allowed in a specified test interval time. Duty cycle is expressed as a percentage of the maximum time that the power source can be operated to its rated output without exceeding the prescribed temperature. A 100% duty cycle power source is designed to provide its rated output continuously without exceeding the specified temperature limit. A 60% duty cycle refers to the power source which can produce its rated output for 6 out of every 10 minutes without getting overheated. Manual welding power sources are designed for 60% duty cycle whereas automatic and semiautomatic welding power sources are rated for 100% duty cycle.

2.2 TYPE OF CURRENT AND POLARITY IN TIG WELDING

TIG welding can be performed using either a DC or AC power source, and the three popular modes used are direct current electrode positive (DCEP), direct current electrode negative (DCEN) and alternating current mode. Each polarity has individual features that make them necessary for specific conditions or certain types of metals. Heat distribution between the tungsten electrode and the weld metal and the degree of surface oxide cleaning determine the current and polarity to be used.

2.2.1 Direct Current Electrode Negative (DCEN)

DCEN is also known as direct current straight polarity (DCSP). In this, the electrode is connected to the negative terminal of the power supply and the workpiece is connected to the positive terminal. Electrons flow from the electrode to the workpiece and positive ions flow from the workpiece to the electrode. During welding, two-thirds of welding heat is concentrated on the workpiece and the remaining one-third at the tungsten electrode. Because of high heat input towards the workpiece, narrow and deep penetration welds are capable of being produced. As one-third of heat energy is generated at the tungsten electrode, a smaller diameter electrode can fulfil the requirement. DCEN is the most commonly used mode in TIG welding to weld most metals.

2.2.2 Direct Current Electrode Positive (DCEP)

DCEP is also known as direct current reverse polarity (DCRP). In this, the electrode is connected to the positive terminal of the power supply, and the workpiece is connected to the negative terminal. Electrons flow from the workpiece to the electrode, and positive ions flow from the electrode to the workpiece. During welding, one-third of welding heat is concentrated on the workpiece and two-thirds is towards the electrode. Because of low heat input at the workpiece, DCEP produces wide and shallow penetration welds. However, it enhances oxide cleaning action upon the base metal. Due to the higher heat input at the tungsten electrode, a large diameter electrode has to be used. The low heat input and strong cleaning action on the workpiece make this mode good for welding thin sheets of aluminium and magnesium which have natural oxide layers on the surface. The workpiece does not emit electrons as freely as tungsten, and so the arc may wander or become more erratic than DCEN.

2.2.3 Alternating Current (AC)

Alternating current (AC) produces a periodic reversal of electrode positive and electrode negative instantaneously. Therefore, AC can effectively combine the deep penetration characteristic of DCEN and the workpiece cleaning action of DCEP. In general, AC is preferred for welding metals such as aluminium and magnesium. DCEN is preferred for welding most other materials and for automatic welding of thick sections of aluminium. When AC is used with argon as the shielding gas, an arc cleaning action is produced at the joint surfaces on aluminium and magnesium. This cleaning action removes oxides and is particularly beneficial in reducing weld porosity during aluminium welding. When using DC, helium may be used as the shielding gas to produce deeper penetration. However, stringent pre-cleaning of aluminium and magnesium parts is required with helium shielding. Argon and helium mixtures for gas shielding can provide the benefits of both the gases.

2.3 ARC CHARACTERISTICS

A welding arc is an electric discharge between two electrodes, and it is characterized by relatively low voltage and high current. An arc is a gaseous electrical conductor that converts electrical energy into heat. The energy produced in the arc per unit time is equal to $V \times I$, where V is the arc voltage and I is the welding current. The welding current is conducted from the electrode to the workpiece through a heated and ionized gas, called plasma, comprising negative electrons and positive ions. The electrons are produced by thermionic emission, and the positive ions are produced by secondary collisions between these electrons and atoms in the gaseous medium to maintain charge neutrality.

The name 'arc' is derived from the shape of hot gas when the electrodes are placed horizontally to each other. Due to its lower density, the hot gas tends to rise, forming an 'arc' shape. In actual welding conditions, the distance between the tungsten electrode and the flat workpiece is about 1–10 mm and the arc assumes a bell-shape. The arc is divided into three separate regions, viz., arc column, cathode fall zone and anode fall zone. The arc column zone occupies the largest space between the electrodes, which is characterized by a small and constant voltage gradient. Adjacent to the electrodes, the zones are called fall zones, characterized by voltage drops: the cathode fall zone in front of the cathode (negative electrode), and the anode fall zone in front of the anode (positive electrode).

The temperature in the welding arc ranges between 5,000 K and 30,000 K, and it depends on the nature of plasma and current conducted by it. For TIG welding, the central core temperature of the plasma can go up to 30,000 K. It may be lowered because of the presence of metal vapour from the non-consumable tungsten electrode and any molten metal particles from the filler used, and the reduction can be attributed to heat losses due to radiation, convection, conduction and diffusion. Based on the type of shielding gas, different temperatures are required to keep the plasma ionized. Argon is easier to ionize than helium due to its low ionization potential. A higher voltage drop and hence a higher heat input to the weld pool is produced when welding is performed using helium or helium mixed gases [den Ouden and Hermans, 2009].

2.4 ARC INITIATION TECHNIQUES

There are several methods available to initiate the arc in TIG welding, including the touch start (also called scratch start or lift start), the high-frequency start, the pulse start and the pilot arc start techniques. The method of arc initiation depends on equipment capability, production requirements and ease of application.

2.4.1 Touch Start

In this method, the torch is lowered towards the workpiece until the electrode makes initial contact with it and subsequently retracted to maintain a short gap to establish the arc. Since it is a simple method, it is used to initiate the arc in both manual and automatic welding processes. The limitation of this technique is that the electrode may stick to the workpiece and may cause tungsten contamination.

2.4.2 High-Frequency Start

High-frequency generators contain a spark-gap oscillator which produces a high-voltage AC output at radio frequencies in the welding circuit. The presence of high voltage ionizes the gas between the electrode and the workpiece and these ionized gases subsequently conduct the welding current to initiate the arc. High-frequency starting is used with either DC or AC power sources for both manual and automatic welding.

2.4.3 Pulse Start

This method is normally used with DC power sources in automatic welding. A high-voltage pulse is applied between the tungsten electrode and the workpiece to ionize the shielding gas and establish the arc.

2.4.4 Pilot Arc Start

In this technique, a pilot arc is established between the welding electrode and the torch nozzle. The pilot arc provides the ionized gas required to create the main welding arc. The pilot arc is generated by an auxiliary power source and is started by high-frequency initiation.

2.5 WELDING EQUIPMENT

In manual TIG welding, the primary equipment comprises a welding power source, welding torch, insulated power cables and shielding gas cylinders. Manual welding torches usually contain auxiliary switches and valves attached to the torch handles to control the current and gas flow. In automatic welding, travel speed control unit, automatic wire feeders and automatic voltage control units are added. In this case, welding torches are mounted on a weld head or carriage and moved along the joint.

2.5.1 Welding Torch

The welding torch is used to grip the tungsten electrode and provides the required shielding gas to the arc and weld pool. They are designed to hold various types and sizes of electrodes and gas nozzles, and consist of various parts such as a tungsten electrode, collet, gas nozzle and gas lens. Torches are designed based on the maximum welding current that can be used without overheating.

2.5.2 Gas-Cooled and Water-Cooled Torches

The heat generated in the torch during welding is cooled by either air or water. Air-cooled or gas-cooled welding torches can be used up to 200 A, whereas water-cooled torches are used for higher current ranges, typically between 300 A and 500 A operating in continuous duty cycles. Water-cooled torches are commonly used in automatic welding applications.

2.5.3 Collet and Collet Body

A collet is a device used to hold the tungsten electrode in the torch. Electrodes of different diameters are kept in position within the torch by appropriately sized collets, also called chucks. Collets and collet bodies are made of copper alloy and a proper contact between the electrode, collet and collet body is necessary for proper current flow and electrode cooling.

2.5.4 Nozzles

Nozzles are made of ceramic materials such as alumina, silicon nitride or fused quartz. Their function is to direct the shielding gas onto the weld pool and their size is determined based on the electrode size, type of weld joint, weld area to be effectively shielded and the access required for welding. The nozzle sometimes has a gas lens to improve the gas flow pattern. It is a part fit around the electrode or collet which contains a porous barrier for providing a laminar flow of shielding gases.

2.6 ELECTRODES AND THEIR CLASSIFICATION

Due to the very high melting point (~3,400°C) and high electrical conductivity, tungsten-based electrodes are used in TIG welding as a non-consumable electrode. The tungsten electrode acts as one of the electric terminals to generate the arc. Due to the initiation of the arc, the tungsten electrode is heated up to higher temperature and subsequently, the electrons are emitted by thermionic emission. The quality of a TIG-welded joint depends on the type of tungsten electrode used and the shape of the electrode tip. Elements such as thorium, cerium and zirconium are added to tungsten to improve the performance of the electrode. The selection of electrode size depends on the welding current, which in turn is dictated by the thickness of the sheets to be welded. Use of higher welding current makes the electrode erode at a higher rate or melt gradually.

Electrodes are classified based on the alloying elements added to the tungsten electrode. Type of electrode, alloying elements added and colour coding are furnished in Table 2.1.

2.6.1 Pure Tungsten, EWP

EWP electrodes are unalloyed tungsten electrodes, suitable for AC welding of aluminium and magnesium alloys. Their current

TABLE 2.1   Different Types of Electrodes for TIG Welding

AWS classification	Colour	Alloying oxide	% oxide
EWP	Green	–	–
EWTh-1	Yellow	Thorium oxide (ThO_2)	0.8–1.2
EWTh-2	Red	Thorium oxide (ThO_2)	1.7–2.2
EWZr-1	Brown	Zirconium oxide (ZrO_2)	0.15–0.4
EWCe-2	Orange	Cerium oxide (CeO_2)	1.8–2.2
EWLa-1	Black	Lanthanum oxide (La_2O_3)	0.8–1.2
EWLa-1.5	Gold	Lanthanum oxide (La_2O_3)	1.3–1.7
EWLa-2	Blue	Lanthanum oxide (La_2O_3)	1.8–2.2

carrying capacity and electron emission characteristics are not as good as those of alloyed tungsten electrodes. EWP electrodes are of low cost and are used in less critical applications.

2.6.2 Thoriated Tungsten, EWTh

EWTh electrodes show better electron emission characteristics and arc stability. They possess a longer life, experience lower electrode tip temperature and show high resistance to tungsten contamination in the weld. EWTh electrodes are suitable to weld steels, stainless steels and nickel alloys in DCEN polarity, and they are not recommended for welding of aluminium and magnesium alloys.

2.6.3 Zirconated Tungsten, EWZr

EWZr electrodes are suitable for welding with AC power sources since they have better resistance against tungsten contamination in the welds. Also, they provide better electron emission characteristics and arc stability. EWZr electrodes are characterized by longer life, lower electrode tip temperature and high resistance to tungsten contamination.

2.6.4 Ceriated Tungsten, EWCe

EWCe electrodes show easy arc initiation, better arc stability and reduced rate of tungsten vaporization. EWTh electrodes are replaced by EWCe electrodes in many countries due to their non-radioactive nature. These electrodes can be used in both AC and DC.

2.6.4.1 Electrode Tip Preparation

Electrode tip configuration is a variable in the TIG welding process and any change in the electrode geometry may affect the arc characteristics, penetration, and the size and shape of the weld bead. Electrode tip configuration is optimized during the development of the welding procedure, and electrode tips are generally prepared by grinding and balling. Silicon carbide grinding wheels are used to get the required shape, whereas balling is achieved by striking an arc on a water-cooled copper block or another suitable material using AC power source or with DCEP mode. In DC welding, thoriated and ceriated electrodes are used in truncated shape with specific included angle. As the electrode included angle increases, the penetration increases and the weld bead width decreases. For AC welding, pure tungsten and zirconated electrodes are used with a hemispherical (ball-ended) shape.

2.7 SHIELDING GAS

Shielding gas is a significant variable in the TIG welding process and it serves two important functions. The shielding gas provides plasma for the conduction of current and it protects the hot metal and the weld pool from interacting with the atmospheric constituents and prevents any adverse reactions. Although the primary function of shielding gas is to protect the weld pool, the characteristics of the shielding gas influence the behaviour of the arc and the resultant weld bead. A chemically inert gas is required for TIG welding to provide shielding of the weld pool. Argon is the preferred choice of shielding gas as it is more easily available in the atmosphere (~0.94% volume). Helium is also used as a shielding gas for welding similar/dissimilar materials. Argon in combination with helium is used to improve the welding characteristics on various materials. Important characteristics of shielding gases are density, ionization potential, thermal conductivity, boiling point, ability to stabilize the arc and ability to ignite/reignite the arc, etc. Inert shielding gases namely argon and helium are used in TIG welding processes while other shielding gases like nitrogen and carbon dioxide mixed in some proportions with argon/helium are also used. The physical properties and characteristics of both argon and helium, which can be used as shielding gases, are compared in Table 2.2. The ionization potential of argon is less than that of helium; hence the ease with which the shielding gas forms plasma is better with argon. The higher the ionization potential, the more difficult is the initiation of the arc. A higher

TABLE 2.2   Characteristics of Shielding Gases: Argon and Helium

Argon	Helium
The ionization potential of argon is 15.7 eV. Therefore, it is easy to ionize the gas and arc initiation is easier.	The ionization potential of helium is 25.4 eV. Therefore, more potential is required to ionize the gas and arc initiation is not as easy as in argon.
The density of argon is 1.784 kg/m³. It is heavier than helium, and therefore provides better shielding.	The density of helium is 0.178 kg/m³, which is 10 times lighter than argon, and therefore shielding is not as effective as argon. For proper shielding, the helium gas is to be operated with higher flow rates.
The arc voltage between the electrode and the workpiece is observed to be low.	The arc voltage produced between electrode and workpiece is higher than argon.
The voltage drop across the arc is less.	The voltage drop across the arc is more. Therefore, higher arc temperature is produced for similar current intensity. Deeper penetration welds can be achieved at higher welding speed.
Better arc stability and arc length control.	Arc stability and arc length control are not as high as for argon.
With DCEP, greater oxide cleaning is witnessed.	The oxide cleaning effect is less than that of argon.
Low cost.	More costly than argon.

ionization potential also leads to poor arc stability and therefore the stability of the arc when using helium as a shielding gas is less. The thermal conductivity of the shielding gas is also one of the important parameters as it pumps in more heat to the workpiece. It is also reported to affect the weld bead shape and a broader weld puddle is usually seen when helium is used as a shielding gas. The density of helium is almost one-tenth of argon and thereby helium requires a larger flow rate to prevent the ingression of oxygen and other contaminants to prevent the contamination of the weld pool.

In general, arc voltage for a given arc length is found to be higher for helium shielding gas. This results in a hotter helium arc than argon arc. Hence, helium is preferred for welding of thick plates at high speed, especially for metal systems having higher thermal conductivity and higher melting points. Helium offers higher thermal conductivity than argon and thereby helium effectively transfers the heat from the arc to the base metal, which in turn helps in increasing the welding speed and arc efficiency [Norrish, 2006].

2.8 FILLER WIRE IN TIG WELDING

Apart from the shielding gases and the electrical power used, the main consumable is the filler wire. The process can be used without the addition of any filler metal, especially for welding thin plates. If a filler is required, then it is added to the weld pool in the form of a rod/wire which can be added either manually or by an automatic wire feed unit. It is also added in the form of a fusible insert to produce an accurate penetration bead through a joint.

2.9 EFFECT OF WELDING PROCESS PARAMETERS

The main process parameters in the TIG welding process are welding current, welding voltage, travel speed, wire feed rate and the nature of shielding gas and its flow rate. In general, welding current is directly proportional to weld penetration and thus thicker plates require larger values of current for achieving deeper penetration. Current has a strong influence on welding voltage. To maintain constant arc length, welding voltage is adjusted proportionally with the change in welding current. Welding current affects the amount of base material melted, filler wire deposition rate and depth of penetration. Excess penetration is observed for higher welding currents and due to the higher heat input associated with it, distortions are also witnessed. Insufficient current leads to a lack of penetration and lack of fusion.

Welding arc voltage is measured between the tungsten electrode and the workpiece when the arc is initiated. Arc voltage is directly proportional to arc length (distance between the tungsten electrode and the workpiece) and it is influenced by welding current, type of shielding gas, ambient pressure and the distance between the tungsten electrode and the workpiece. Arc length affects the width of the weld pool (pool width is proportional to arc length) and to some extent, the penetration and shielding. Variables such as the type of electrode, shielding gas, filler wire feed position, temperature changes in electrode and electrode erosion affect the arc voltage and thereby the arc length is changed. Arc voltage control (AVC) systems are adapted in automatic welding to adjust the arc length and restore the required arc voltage.

The weld travel speed is also found to affect the weld performance and it is defined as the linear rate at which the arc moves with respect to the plate along the weld joint. It is inversely proportional to weld penetration and the width of the weld. However, the effect of travel speed on width is more noticeable than its effect on penetration. Higher welding speed results in low heat input and low filler metal deposition rate and promotes the chances for defects such as undercut, porosity, uneven bead shape and arc blow. A lower travel speed pumps in more heat to the base metal and thereby the width of the weld increases. The filler deposition rates are also to be controlled to match the welding speeds. In automatic welding, the speed is kept constant while other variables, such as current and voltage, are varied for obtaining sound weld joints.

Wire feed rate is defined as the amount of filler wire or length of filler wire deposited per unit of time. In manual TIG welding, the amount of filler metal to be fed and the depositing technique are controlled by the welder. Wire feed rate and the feeding technique are carefully adjusted to reduce the incomplete fusion and to achieve a better bead appearance. In automatic welding, the wire feed rate is predetermined and fixed in the auto wire feeder which deposits the filler wire at a specified speed. Weld penetration increases with a lower wire feed rate and makes the weld bead contour flatten. A very low wire feed rate leads to defects such as undercut, incomplete joint fill and solidification cracking and a higher wire feed rate produces a more convex weld bead and decreases the weld penetration.

Shielding gas flow rate is a critical parameter in TIG welding and is generally represented in litres per minute or cubic feet per hour. An adequate amount of gas is required to shield the weld pool and the tungsten electrode to prevent any adverse reactions with the atmospheric constituents. Insufficient gas flow leads to porosities or other entrapments and excessive gas flow is a waste of shielding gas. Gas flow rate is increased with an increase in welding current and nozzle diameter. During welding in AC mode, 25% high flow rate is required as the current reversals cause a disturbing effect on shielding gas.

2.10 DISCONTINUITIES AND DEFECTS

Discontinuity is an interruption in the typical structure of a material, such as a lack of homogeneity in its mechanical, metallurgical or physical characteristics. A discontinuity is not necessarily a defect. If a discontinuity exceeds the acceptance criteria, it is classified as a defect and is either repaired or rejected. Defects are discontinuities that occur by nature or an accumulated effect (e.g., total crack length), which limit the weld joint' ability to meet the minimum requirements intended. Defects are the regions where the metal is absent (e.g., porosities, cracks), regions where there are low-density non-metallic inclusions (e.g., slag or flux entrapment), regions where there are high-density inclusions (e.g., tungsten inclusions) or regions of various geometric discontinuities (e.g., lack of penetration, mismatch or undercut). Defects in welds occur due to one or more of the following reasons: (1) improper joint design, preparation and fit-up, (2) base or filler metal intrinsic characteristics, (3) process features and (4) environmental factors. Irrespective of origin, defects always act as regions of stress concentration and degrade the mechanical properties and performance of the weldments [O'Brien, 2004].

2.10.1 Porosity

During welding, gases such as hydrogen, oxygen and nitrogen get entrapped inside the weld pool and form cavity-type discontinuities called porosities. Hydrogen is one of the important sources for causing porosities. The solubility of dissolved gases decreases with a decrease in temperature and if these gases are present at more than their solubility limits, they are expelled during solidification in the form of gas bubbles or pockets.

2.10.1.1 Causes

- Excess entrapped gases such as hydrogen, oxygen and nitrogen in the welding atmosphere.

- Contaminated base metal and filler metal and presence of impurities in shielding gas.

- Improper welding current and arc length.

- High solidification rate.

2.10.1.2 Remedies

- Use clean base metal, filler metal and high-purity shielding gases.

- Preheat the base metal and increase the heat input to reduce the solidification rate.

- Use of filler metal that contains deoxidizers.

- Ensure proper welding techniques and conditions.

2.10.2 Lack of Penetration

A defect that occurs at the root of the joint when the weld metal does not reach through the joint thickness or the weld metal fails to fuse completely with the root faces of the joint.

2.10.2.1 Causes

- Insufficient welding current.

- Excessive travel speed.

- Improper joint preparation.

2.10.2.2 Remedies

- Increase the welding current.

- Reduce travel speed.

- Proper joint design to access the bottom of the joint.

2.10.3 Lack of Fusion

A weld discontinuity that occurs due to a lack of coalescence between the weld metal and the base metal or between the weld metal and an adjacent weld bead. If the lack of coalescence is between weld metal and base metal, it is called lack of side wall fusion and if the lack of coalescence is in between the weld beads in multi-pass welds, it is called as an inter-pass cold lap.

2.10.3.1 Causes

- If the base metal surfaces or previously deposited weld metals contain excessive oxides, the chances of fusion are meagre.

- Improper welding technique.

- Improper joint design and edge preparation.

2.10.3.2 Remedies

- Clean all groove faces and weld zone surfaces before welding.

- Maintain optimum root face and root opening.

- Adopt high welding current.

- Use a smaller travel speed.

2.10.4 Undercut

A groove melted into the base metal adjacent to the weld toe or weld root and unfilled by weld metal. Undercutting produces a weak region in the weld and if it exceeds the acceptance criteria it is considered a defect and must be repaired.

2.10.4.1 Causes

- Excessive welding current.

- Improper electrode size.

- Mismatch between electrode design and weld position.

2.10.4.2 Remedies

- Use moderate welding current and correct welding speed.

- Proper weaving technique.

- Use an appropriate electrode size that produces a proper puddle size.

- Proper positioning of the electrode relative to a horizontal fillet weld.

2.10.5 Tungsten Contamination of Workpiece

A defect that occurs due to entrapment of tungsten particles from the electrode which form solid inclusions in the weld metal.

2.10.5.1 Causes

- Touch starting with the electrode.

- Electrode melting and fusing with the base metal.

- Contact of tungsten electrode with the weld pool.

2.10.5.2 Remedies

- Tungsten inclusions can be controlled by use of thoriated or zirconiated tungsten electrodes in place of pure tungsten.

- Use high-frequency starter; use copper striker plate.

- Use less current or larger diameter electrode; use proper electrode for the material being welded.

- Keep tungsten out of the weld pool.

2.10.6 Crack

A fracture-type discontinuity characterized by a sharp tip and high ratio of length and width to opening displacement. A crack which occurs during the weld solidification is called hot cracking whereas one which develops after solidification is called cold cracking or delayed cracking. Hot cracking susceptibility mainly depends on the chemical composition of solidifying weld metal, restraint during welding, development of thermal stresses, etc. The severity of cold cracking depends on susceptible microstructure, residual stresses and amount of elemental hydrogen, etc.

2.10.6.1 Causes

- High restraint weldment.

- Improper selection of filler metal.

- Improper joint preparation.

- High amount of impurity elements (S, P, B, etc.).

- Undesirable microstructure.

2.10.6.2 Remedies

- Right selection of filler metal.

- Design the structure and develop a welding procedure to eliminate high restraint joints.

- Employ preheating and inter-pass heating.

- Ensure sufficient ductility in the weldments.

2.10.7 Distortion

The difference in the expansion of weld metal and adjacent base metal during the welding process tends to bend/twist the workpiece. Distortion can be classified as longitudinal shrinkage, transverse shrinkage and angular distortion.

2.10.7.1 Causes

- Larger weld volume.
- Improper welding sequence.

2.10.7.2 Remedies

- Reduce the volume of weld metal.
- Introduce double-V joint and welding alternately on either side of the joint.
- Placing welds around the neutral axis.
- Plan the welding in a proper sequence.
- Reduce number of passes in multi-pass welding.

2.10.8 Overlap

Overlap occurs when the weld face extends beyond the weld toe. In this condition the weld metal rolls and forms an angle less than 90 degrees.

2.10.8.1 Causes

- Improper welding technique.
- Use of large diameter electrodes.
- High welding current.

2.10.8.2 Remedies

- Adapting a proper welding technique.

- Use small diameter electrode.

- Less welding current.

2.11 ADVANCES AND MODIFICATIONS IN TIG WELDING

2.11.1 Pulsed Gas Tungsten Arc Welding (P-GTAW)

A cyclic variation in direct current (DC) from a low background value to a high peak value results in pulsed gas tungsten arc welding (P-GTAW). Pulsed DC power sources permit modifications in pulse current time, peak current level, background current time and background current level to achieve the required current output. Peak current helps in achieving good fusion and deeper penetration while the background current helps in sustaining the arc and allows the weld to solidify. Advantages of P-GTAW compared to continuous current welding are good penetration with less heat input, minimized distortion, good control of the pool, easy to weld thin materials and material with dissimilar thickness.

Another variation in P-GTAW is high-frequency pulsed current welding. In regular pulsed current welding, a frequency range of 0.5–20 pulses per second is achieved. In high-frequency mode, 200–500 pulses per second is achieved and thus it produces a stiffer arc (less arc wandering). Also, it promotes arc agitation in the weld pool, which helps the impurities to float to the surface, resulting in a weld with better metallurgical properties. High-frequency pulsing is used in precision and automatic welding where an arc with superior directional properties and stability at lower current is required. Also, pulsing mode interrupts the solidifying weld metal and breaks the dendritic arms, thereby resulting in a fine-grained microstructure.

2.11.2 Hot Wire TIG

Hot wire TIG is adapted to increase the deposition rate and the productivity of the process. Filler wire is resistance heated to its near melting point before reaching the weld pool. As the filler wire is preheated, proper shielding is required to protect the hot wire. Due to the preheating of the filler wire, most of the arc energy is available for melting the base metal. Compared to cold wire feeding, this process has a higher deposition rate and is capable of welding with a higher welding speed.

2.11.3 Arc Voltage Control (AVC)

AVC is used in automatic gas tungsten arc welding to maintain arc length. In this method, the arc length is continuously monitored and any variation in the arc length is adjusted automatically using a feedback system. If the arc length is changed due to electrode movement, the desired voltage is automatically adjusted by sensing the arc length.

2.11.4 Gas Tungsten Constricted Arc Welding (GTCAW)

This method is also known as the InterPulse welding technique developed by VBC Group, UK. InterPulse operates at very high frequency (20,000 Hz) and produces a magnetically constricted columnar profiled arc similar to the plasma arc, which significantly reduces the overall heat input during welding. Due to very low heat input, the cross-sectional area of the weld is reduced. The technique was originally developed to weld nickel-based superalloys and was subsequently extended to titanium alloys. This technique is also recommended to repair the un-weldable and precision superalloy castings and produces a narrow weld bead and heat-affected zone. Due to a focused arc and low heat input in this technique, titanium alloys can be successfully welded in the shop floor instead of welding in costly and capital-intensive welding chambers and trailing shields.

2.11.5 Arc Oscillation

Oscillation of the arc helps in covering a larger area of the weld metal and thus increases the productivity. It can be done either mechanically or using any magnetic sources. In mechanical arc oscillation, an alternating (side-to-side) motion relative to the direction of travel is used to increase the width of the welds. Mechanical oscillation can be achieved by mounting a TIG torch on a cross slide that provides automatically controlled movement of the torch and wire feed transverse to the line of travel. This equipment provides adjustable cross-feed speed, variable oscillation amplitudes and change in dwell time on either side of the oscillation cycle.

Magnetic oscillation can also be used to increase the width of welds. Magnetic control can be used to reduce the disruptive effects of arc blow. A four-pole magnetic probe is used to reduce arc deflection and maintain the arc position. Oscillators deflect the arc longitudinally or laterally over the weld pool without moving the welding electrode or wire feed. The oscillators consist of electromagnets located close to the arc, powered by a variable-polarity, variable-amplitude power source.

2.12 ADVANTAGES AND LIMITATIONS OF TIG WELDING

The following are some of the advantages of the TIG welding process:

- Produces extremely high-quality welds, generally free of defects, using cheaper power sources.

- Generally, spatter-free welds are produced.

- Filler wire addition depends on the requirement.

- Allows excellent control of root pass weld penetration.

- Produce inexpensive autogenous welds at high speeds.

- Welding process parameters can be controlled precisely.

- Used to weld almost all metals, including dissimilar metal joints.

- Allows the heat source and filler metal additions to be controlled independently.

The following are some of the limitations of the TIG welding process:

- Deposition rates are lower than the rates possible with consumable electrode arc welding processes.

- Welder skill is mandatory in this process as a proper coordination between torch movement and filler metal feeding is essential.

- This process is considered to be less economical than other arc welding processes for welding thicker sections.

- There is difficulty in shielding the weld zone properly in drifty environments.

2.12.1 Potential Problems with the Process

- If the tungsten electrode makes contact with the weld pool, defects like tungsten inclusions may occur.

- Contamination of the weld metal can occur if proper shielding of the filler metal by the gas stream is not maintained.

- There is low tolerance for contaminants on filler or base metals.

- Improper cleaning of base metal and filler metal, impurities in the shielding gases and coolant leakage in water-cooled torches can cause porosities in the weld.

2.13 SAFETY PRECAUTIONS

The major safety requirements in TIG welding are proper insulation of power cables, ensuring safe fixing of return clamps and safety against welding fumes and arc radiations. The high-frequency spark used to initiate the arc can occur through any break in the insulation and can cause deep burns which can be very painful. The delivery of inert gases such as argon and helium into confined spaces may lead to the oxygen content of the air being reduced to a point where there is a danger of asphyxiation. Enough ventilation during welding is essential to avoid this situation. Advanced in-situ fume extraction hoods are available to remove toxic fumes generated during welding. For safe practice, the welder should be equipped with personal protective equipment such as welding helmet with appropriate shade, suitable respiratory mask, flexible welding gloves, apron and welding goggles.

CHAPTER 3

Activated Tungsten Inert Gas (ATIG) Welding

THE GAS TUNGSTEN ARC welding process is most suited for welding operations that require joints to have considerable precision with high joint quality. Tungsten inert gas (TIG) welding dominates critical industries like aerospace, nuclear, etc. as it produces high-quality welds. However, usage of the TIG welding process is restricted due to its low deposition rate and limited penetration capability. The advantages of TIG welding are offset by its disadvantages, such as the capability of the process to weld only a limited thickness metal in one single pass and the poor productivity of the process in welding thicker sections. The poor productivity is a result of low welding speeds and the requirement of multiple passes for welding thicker sections.

The concept of using an activating flux was first proposed by the E. O. Paton Institute of Electric Welding, Ukraine. In 1965, the institute developed a modified TIG welding process to increase the penetration: the activated tungsten inert gas (ATIG) welding process or

Paton activated TIG (PA-TIG) welding. ATIG welding uses a thin layer of an activating flux coating on the surface of the metal to be welded to overcome these limitations. A significant increase in the depth of penetration is achieved for given welding conditions. This improved technique extends the operating range of the TIG welding process by the use of fluxes. These fluxes reduce the requirement of edge preparations and increase productivity due to a decrease in the number of passes needed to produce the joint. Further, the ATIG process can autogenously weld a thickness up to 10 mm in a single pass using square butt edge preparation. It can be used as either a manual or automatic welding process. ATIG welding is a modification of the TIG welding process. It does not require any special attachment or equipment to increase the weld penetration unlike high-energy beam welding processes like plasma arc welding and laser beam welding. Consequently, the ATIG process was expected to increase productivity for the industries.

A layer of an activating flux paste is applied over the surface to be welded (see Figure 4.1). The flux is usually in the form of powders and mixed with any solvent or a binder to form a paste and can be either brushed or sprayed on the surface of the material to be welded. The activating fluxes are predominantly oxides of metals and in a few cases, halides are also used. However, there can be separate formulations of fluxes for welding of a specific metal. Fluxes have been developed for materials like C-Mn steels, low alloy steels, Cr-Mo stainless steels, nickel-based alloys and titanium alloys. During welding, activating flux evaporates and constricts the arc column, which increases the arc energy density. So, weld penetration increases significantly compared to conventional TIG welding. The Paton Welding Institute (PWI) of Ukraine and the Edison Welding Institute (EWI) of the USA developed fluxes and made them commercially available. However, the composition of fluxes is patent protected. The composition of the fluxes varies based on the material and application. Generally, oxides of titanium (TiO_2), silicon (SiO_2), chromium (Cr_2O_3) and nickel (NiO) are used as fluxes and in some cases, a small addition of halides like sodium fluoride (NaF), calcium fluoride (CaF_2) and aluminium fluoride (AlF_3) is also found to be effective. It is also observed that the use of flux does not produce any significant change in the chemical composition of the weld. Also, they do not degrade the mechanical properties and microstructure of weldments; rather they improve the mechanical properties of welds compared to conventional TIG welding process. However, if there was any entrapment of flux particles in the weldment, the mechanical properties would be affected drastically. To ensure the weldment is free from such contaminations, the process parameters are to be optimized with the right choice of flux for the selected material to be welded.

The precise mechanism for the process improvement is not fully determined. However, theories have been proposed which state that the surface fluxes influence the surface tension of the liquid metal or release ionic species, which enable the arc constriction. During welding, the flux is evaporated and constricts the arc by capturing electrons in the exterior regions of the arc. Towards the arc centre, high temperatures prevail and ionization dominates due to the presence of a strong electric field. In the periphery of the arc, the temperature is lower than in the central region and the electrons have low energy in a weak electric field. Therefore, electron attachment occurs in the low-temperature peripheral regions. This phenomenon restricts the current flow to the central region of the arc and increases the current density in the plasma leading to a narrower arc and deeper weld. An increase in the dissolved oxygen content changes the coefficient of surface tension from negative to a positive value and favours reverse Marangoni flow (detailed explanation in Chapter 4). Therefore, a narrow and deep weld is produced. The combination of these two mechanisms leads to increased penetration in ATIG welding.

Marangoni movements are centrifugal in TIG welding and they do not facilitate the weld penetration. In ATIG, the presence of surface active elements changes the temperature coefficient of surface tension from positive to negative, thereby reversing the Marangoni flow resulting in higher depth of penetration. While using oxides as activating fluxes, the main factors influencing the depth of penetration in ATIG welding are the amount of oxygen in the welding zone, surface tension and its temperature gradient, as well as its intensity and direction of molten metal flow in the weld pool.

Various phenomena that occur during ATIG welding can be divided into two zones, viz. the arc (zone A) and the weld pool (zone B). In zone A, energy, heat, geometry and force parameters of the arc are characterized. Mechanisms associated with arc constriction are studied in zone A. In zone B, the size and profile of the weld pool, its surface tension and temperature and the chemical composition of the welding pool are characterized. The changes in size, profile, surface tension and temperature occurring in the weld pool, as well as the formation and intensification of molten metal flow in the weld pool including the Marangoni flows, are studied in zone B.

The ATIG process claims several advantages over the conventional TIG welding process. The advantages are listed below:

a) Improves weld penetration, and therefore a thicker plate can be welded in a single pass.

b) No edge preparation (bevelling or groove preparation) is required to weld thicker plates in one single pass.

c) Since the number of passes is reduced, the applied heat input is less. Therefore, heat input-related weldability issues are minimized.

d) Weld residual stresses and weld distortion are minimized.

e) Filler wire requirement is reduced as weld volume is lower.

f) Reduces welding time as much as 50% and increases productivity.

g) Heat input, weld time and power and labour requirements are reduced; thereby manufacturing cost is also reduced.

ATIG welding has potential applications in the nuclear, aerospace, defence and chemical industries, where thick sections of pressure vessels, pipes and tubes are fabricated. Sometimes, this process can be used for repair welding where deeper penetration is required. The ATIG process can also be used for welding thin sections with the condition that welds are to be made at higher speeds. It ensures minimal heat input to the material compared to the conventional TIG process and thereby minimizes heat-associated damage.

However, it has been observed that, when ATIG welding is performed on materials having low melting temperature and boiling temperatures with high thermal conductivity, the resulting temperature in the weld pool is also lower. This leads to a poor temperature gradient in the weld pool, which may not result in reverse Marangoni flows. It has also been observed that, during ATIG welding, the flux layer reacts with the welding arc and deposits a slag layer on the surface of the weld. It makes the weld appearance poor and unattractive [Huang et al., 2007].

To make the weld surface lustrous, it is necessary to clean the surface after welding. It is an added effort that often has to be done manually. In a mass production environment, the time and cost of applying activating flux before welding as well as cleaning after welding are major drawbacks. The issue of not being able to maintain weld penetration and weld appearance simultaneously was the motivation for introducing another type of flux-assisted welding, known as flux zoned tungsten inert gas (FZTIG) welding. In this process, instead of a single flux, two different fluxes (flux 1 and flux 2) are used, called central and side fluxes (see Figure 4.1). The central region of the weld, which comes into direct contact with the welding arc, is coated with the central flux, which should have a low melting temperature, low boiling temperature and low current resistivity. The side regions of the weld surface are coated with the side flux, which should have a high melting temperature, high boiling temperature and high current resistivity. After applying fluxes, welding is carried out similarly to ATIG welding. It is observed that the differential application of fluxes causes an increase in weld penetration and gives a clean, smooth weld at the same time.

Huang et al. (2012) studied the effect of welding current on FZTIG welding.

(a) *Welding current*: It was found that, relative to the penetration in TIG welding, FZTIG welding had a higher penetration up to a threshold value of current after which TIG welding gave better results. This happens because, at low values of current, only the central low melting flux evaporates, helping to constrict the arc. At higher current values, the side coat also melts, which makes the arc wide, leading to a wider weld bead.

(b) *Central flux coat width*: As the width of the central flux coat increases, the effect of the constriction of arc brought about by the side flux decreases. For very small or very large widths, the effect is not noticed at all, and the process becomes similar to ATIG welding. So, an appropriate thickness of the central coat should be used to maintain high penetration.

(c) *Central flux coat thickness*: Similar to the flux width, the flux thickness also decreases the penetration of the weld bead if beyond a threshold limit. This happens because the thick coat makes it difficult for the arc to cut through and melt the base metal.

FZTIG welding was specifically devised to tackle the problem of insufficient reverse Marangoni currents in low melting alloys. The process has not been deeply investigated and has a future scope in terms of analysing the interaction of each flux component with the arc and with each other, to study the mechanisms involved and its difference with TIG and ATIG welding, etc.

Another variant of the ATIG process is the advanced activated tungsten inert gas (AATIG) welding process wherein the amount of oxygen/surface active element required for reversal of Marangoni convection currents is supplied along with the shielding gas to enhance the depth of penetration.

CHAPTER 4

Flux Bounded Tungsten Inert Gas Welding (FBTIG)

4.1 OVERVIEW

A more precise weld at a lower power rating, welding of thinner metal sections and the possibility of joining exotic metals are the advantages of tungsten inert gas (TIG) welding over other methods. The lower depth of penetration achieved through TIG welding is one of the key reasons TIG welding is not selected for joining thicker sections of metals/alloys [Messler, 2008]. If TIG welding is resorted to for welding thick sections of metals, multiple passes of the welding torch over the edges to be joined will be required to ensure complete penetration and, moreover, the edges need to be bevelled to suit a particular configuration. The higher heat input, and extra consumption of materials and surface preparations lead to additional cost, reduced productivity and larger heat-affected zones (HAZs) [Brungraber and Nelson, 1973]. Higher heat input during welding will result in differential heating of the base metal leading to distortion, warpage and other associated defects in the welded sections. Thereby, precautions such as post-weld heat treatment, pre-heating, etc. may be required to improve the performance of the welded structure in the intended application if a conventional TIG welding method is adopted.

The Paton Electric Welding Institute of National Academy of Sciences, Ukraine, developed a flux-assisted welding process, namely activated tungsten inert gas (ATIG) welding in the 1960s, to overcome the disadvantages of shallow depth of penetration achieved by conventional TIG welding process [Paskell et al., 1997]. In ATIG welding, a suitable flux coating (generally oxides, fluorides or chlorides) is applied over the weld joint to achieve/improve the required depth of penetration through different mechanisms [Modenesi et al., 2000; Leconte et al., 2013]. Therefore, the number of passes required to join thicker metal sections is found to be less for the ATIG welding process. The weld bead geometry is generally characterized by the depth-to-width ratio (DWR) of the fusion zone, which is defined as the ratio of the depth of penetration to the width of the weld bead. The possibility of the flux mixing with the weld pool and getting entrapped upon solidification is also higher in the ATIG welding process, causing the weld joints to perform more poorly in service. The presence of inclusions or flux particles in the weldments leads to decreased toughness of the weld joint and possibly also brittleness and premature failure. The line of vision of the operator on the edges to be joined is hindered by the flux cover, and thus, the process demands a skilled operator to traverse the torch in the desired manner to trace the edges to be welded. This warrants the automation of ATIG welding process and hence results in a less economical process albeit improving the depth of penetration. Though the ATIG welding process is well established and recognized by the scientific community for having a larger depth of penetration, it could not be carried forward to be adopted and practised in the production industries. The necessity of automation and the difficulty in controlling the welding arc have hindered commercialization of this process. These drawbacks of ATIG led to the advent of flux bounded tungsten inert gas (FBTIG) welding, which alleviates the disadvantages while retaining the main advantage of ATIG: the larger depth of penetration.

Sire and Marya (2002) proposed a new variant of the flux-assisted TIG welding process called FBTIG welding on the pretext of reaping the advantages of ATIG while overcoming its drawbacks. Since flux bounds the arc and surrounds the weld zone, it is called a 'flux bounded' technique. It is also observed that the welding arc follows a different profile based on the application of selective flux and the flux gap. For the corresponding heat input as that to the weld plates, FBTIG welding is observed to provide a larger depth of penetration and also a larger DWR compared to other flux-assisted welding methods such as TIG/ATIG/FZTIG. On the other hand, it can also be inferred that to achieve a similar depth of penetration, the FBTIG process will require less heat input than the other TIG welding methods. This is especially advantageous to prevent the loss of low-melting alloying elements due to excess vaporization. Even though the basic concepts of FBTIG are comparable with ATIG/FZTIG in selective flux application, it can be observed from Figure 4.1d that the edges to be joined are not completely masked by the flux and uncover a small narrow strip of the metal to be exposed to the welding arc, referred to as the 'flux gap'. Thus, the FBTIG welding process can be considered an advancement or variant of the ATIG process in which the welding arc becomes confined to the flux gap. When the flux gap in a typical FBTIG process tends to zero (i.e., the edges to be welded are completely masked with the activating flux), the process becomes the ATIG process (Figure 4.1b), and when the flux gap is large enough to have no influence on activating flux, the process becomes the conventional TIG welding (Figure 4.1a).

Like any other arc welding process, the major parameters for FBTIG welding include welding speed, welding current, arc length, nature of the base metal, etc., and in addition to these basic welding parameters, flux gap, nature of flux, flux particle size and coating thickness play a very significant role in

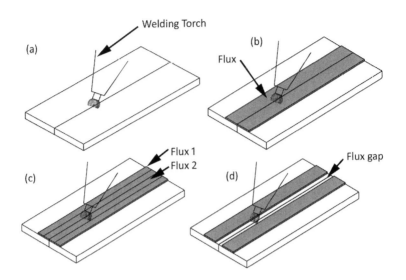

FIGURE 4.1 Various TIG welding processes: (a) conventional TIG, (b) ATIG, (c) FZTIG (d) FBTIG.

achieving the desired weld quality. Understanding the influence of each parameter and their interactions is highly essential for ensuring high-quality welds and thereby to achieve reliable joint performance. Flux gap, flux powder particle size, coating density, etc. are the typical parameters specific to the FBTIG welding method.

4.2 MECHANISMS OF FBTIG WELDING PROCESS

The use of activating flux over the edges to be joined causes the welding arc to be confined in the narrow flux gap and thereby results in increased heat density at the joint. The usual range of penetration for flux-assisted weld methods varies between 3 mm and 7 mm per pass depending on the other parameters, whereas the optimization of parameters can lead to larger depths in one single pass. Numerous investigations into this technique converge on three major reasons for achieving larger penetrations: the insulation effect, reverse Marangoni convection currents and arc root constriction.

4.2.1 Insulation Effect

During welding, electrons/positively charged species are bombarded onto a metallic substrate (workpiece/anode) which converts their kinetic energy into heat energy and causes the metal to melt. Insulating metal oxides or inorganic compounds are mostly used as fluxes and usually have a higher electrical resistance than the base metal. The high electrical resistance offered by the flux shapes the charged species to be channelled to the flux gap, thereby making them bombard directly on the metallic substrate to follow the least resistant path for the conduction of charged species [Zhao et al., 2011]. Figure 4.2a depicts the profile of the welding arc in the absence of flux coating, which results in a diverged (bell-shaped) profile bulging towards the end. In such a case, the heat density input to the base metal will be less than the focused arc. In FBTIG welding, the flux is selectively applied near the edges to form a narrow flux gap and the arc tends to cave in near the base metal so that it becomes channelled to strike the workpiece within the flux gap only. Figure 4.2b shows the focused welding arc that borders the base metal as in the case of the FBTIG welding process. The arc diameter at its tip decreases and the focused arc is supposed to have a higher current density, which increases the depth of penetration. This phenomenon, which is commonly referred to as insulation effect, was identified as one of the prime reasons for enhanced depth of penetration in the FBTIG welding process [Vilarinho et al., 2009; Lowke et al., 2004, 2005]. The insulation

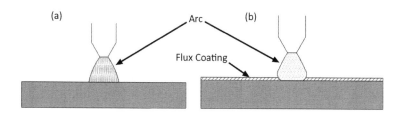

FIGURE 4.2   Typical arc profile in (a) conventional TIG welding process, (b) FBTIG welding process.

effect is exclusive to the FBTIG welding process and it arises due to the nature of the flux application method, which leaves a small narrow flux gap. This effect is seen in conjunction with the other two mechanisms of flux-assisted welding processes as described in the next section. It is also observed from the weld-pool disturbances that the flux layer blocks the spread of the weld-pool front in the width direction, thereby causing the weld-pool length to increase, keeping the weld bead width to a minimum. For the same welding current, a narrow flux gap would generate a welding arc of higher current density and hence, flux gap is one such significant parameter in enhancing the depth of penetration and DWR.

Figure 4.3 shows the profile of the welding arc captured during the welding process without flux coating and in the presence of flux coating (FBTIG), respectively. The flaring of the arc at its bottom can be witnessed from Figure 4.3a as in the case of conventional welding. From Figure 4.3b it is evident that the arc is confined to the flux gap alone on the base metal in FBTIG welding [Jayakrishnan et al., 2017]. Since the flux (SiO_2 in this case)

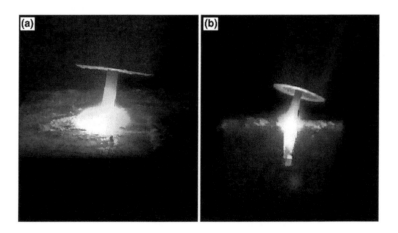

FIGURE 4.3 Welding arc profile during (a) conventional TIG welding, (b) FBTIG welding process for a flux gap of 3 mm with silica as the flux (reprinted from Jayakrishnan, S. et al., *Transactions of Indian Institute of Metals* (2017) 70: 1332, with permission from Springer).

offers very high electrical resistance compared to the base metal, it acts as a barrier and restricts the arc to the flux gap, assuming a shorter diameter towards the base metal. The reduction in the arc diameter results in higher current density and thereby higher penetrations.

4.2.2 Marangoni and Reverse Marangoni Flows

In the 19th century, Italian physicist Carlo Marangoni investigated the effects of surface tension-driven flows within a fluid. The process named after him, the 'Marangoni effect', is defined as the fluid flow that occurs along the interface of two fluids due to the difference in the values of surface tension (γ). The phenomenon, also called the Gibbs–Marangoni effect, can be explained as the result of the presence of a gradient in surface tension that naturally drives the liquid to flow from regions of lower surface tension to regions of higher surface tension. This was initially discovered by physicist James Thompson when he noticed the thin layers of wine climbing the glass column against gravity and dripping back down, which is popularly known as 'tears of wine'. Though he was the first person to discover the phenomenon, the scientific explanation was given by Carlo Marangoni through his doctoral work and explained that fluid always flows in the direction of increasing surface tension. It is also observed that if a small quantity of fluid with low surface tension is poured onto a fluid possessing high surface tension, the flow is set up in the outward direction in such a way that the high surface tension fluid pulls the low surface tension fluid to itself. During any arc welding process, the weld pool also experiences different domains of fluid with different surface tensions and the gradient in the surface tension can be caused either by variation in the composition (concentration gradient) or by a temperature gradient (surface tension is a function of temperature) [Modenesi, 2015]. If the surface tension gradient is caused only due to the difference in the values of temperature within a fluid domain, such a flow is called thermocapillary convection.

The final weld bead shape in a typical fusion welding process is influenced by the weld-pool fluid dynamics, which in turn is due to various other forces acting on the weld pool. In the case of an arc welding process, the temperature at the centre of the arc column is higher and obviously, the temperature at the weld centre line would also be higher. The decreasing trend of temperature is witnessed from the weld centre line towards the periphery of the weld pool. It is also very apparent that the surface tension follows an inverse relation with the temperature and therefore the values of surface tension will be less in the centre line of the weld pool and gradually decrease towards the periphery. In the case of a TIG welding process also, the centre of the weld pool marks the highest temperature due to its closeness to the welding torch and hence the surface tension of the molten metal remains low compared to the other regions of the weld pool. At regions of lower temperature, the values of surface tension remain high. A surface tension gradient is observed within the weld pool and hence Marangoni flow occurs, which sets the fluid in motion from the centre to the periphery (from lower to higher surface tension) [Lu et al., 2004, 2008; Tsai and Kou, 1989]. Figure 4.4a depicts the variation of temperature and surface tension along the weld pool. The temperature along the weld centre line is higher and gradually decreases towards the periphery, whereas the surface tension is lowest at the weld centre line and gradually increases towards the periphery. The variation of surface tension along the weld distance (S) from the centre line of the weld can be mathematically written as $(d\gamma/dS)$.

Rewriting the mathematical representation of surface tension with the distance $(d\gamma/dS)$ as $(d\gamma/dT)\times(dT/dS)$, the variation of γ with respect to temperature and the variation of temperature with respect to distance are well known. Figure 4.4b,c show the depiction of the variations of surface tension with temperature and variation of temperature with respect to distance respectively. Both the curves have a negative slope and thus the resultant, which is the product of $(d\gamma/dT)\times(dT/dS)$, has a positive slope

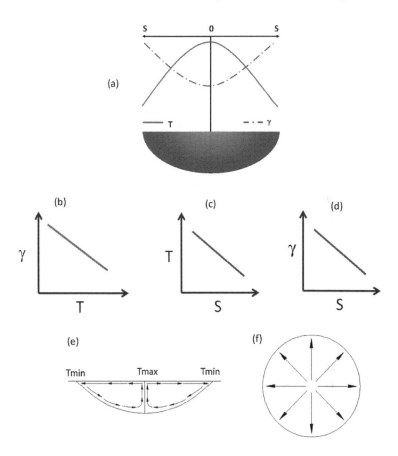

FIGURE 4.4 (a) Variation of surface tension with temperature in the weld pool, (b) variation of γ with temperature, (c) variation of temperature with distance, (d) variation of γ with distance, (e) and (f) depiction of Marangoni flow in the front and top views.

(Figure 4.4d). The weld centre line has the least surface tension and increases as it moves towards the periphery. In this case, the hot fluid from the centre flows outwards towards the periphery and strikes the boundary, causing it to melt and expand in the width direction (Figure 4.4e,f). This results in the weld bead having a larger width and lesser depth, effectively a low DWR. Such an outward flow results in shallow penetration and for welding thicker sections, multi-pass welding becomes mandatory for achieving complete penetration.

As in the case of a flux-assisted welding process (ATIG/FBTIG/ FZTIG), the presence of surface-active elements (SAEs) in the weld pool results in an inward flow, i.e., from the periphery to the centre. It is also observed that certain elements such as oxygen and sulphur segregate at the surface of the metal and block active sites for chemical reactions. Such elements segregate towards the surface to reduce the energy of the system [Seetharaman et al., 2014]. Therefore, these elements retard the rate of reaction and are hence called surface poisoners. Studies of Heiple et al. (1981, 1982, 1983) also indicate that the surface-active elements such as oxygen, and sulphur, etc. influence the direction of molten metal flow in the weld pool during the welding process. The presence of these elements affects the variation of surface tension values with respect to temperature, $\partial \gamma / \partial T$, and changes the gradient from negative to positive as depicted in Figure 4.5a. It can be understood that as the temperature decreases from the centre of the weld pool to its periphery, the value of surface tension also decreases and results in inward convection currents [Heiple and Roper, 1981].

Now, the question that should be asked is, how can the relation of surface tension with temperature be reversed? The perception with which the above discussion lies is with respect to the weld

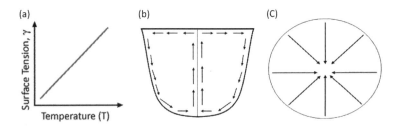

FIGURE 4.5 (a) Influence of activating fluxes on the variation of surface tension with temperature, (b) front view of the weld pool depicting the flow pattern and (c) top view of the reverse Marangoni convection causing inward flows.

centre line and towards the periphery; it is observed that both the temperature and surface tension are decreased. At higher temperatures, the fluid will have lower values of surface tension and the same fluid is expected to have higher surface tension values at lower temperatures. But as depicted in Figure 4.5a, γ increases with an increase in temperature, which shows a direct relationship of γ with temperature (detailed explanation in Section 4.2.3). Similarly, the nature of variation of surface tension with distance from the centre line of the weld pool changes and the fluid at the centre line of the weld pool possesses higher temperature as well as higher values of surface tension. This causes a reversal of convectional currents (the weld-pool convection mode), is termed reverse Marangoni convection and influences the weld penetration. In this particular case, the fluid from the periphery flows towards the centre (regions of higher surface tension) and picks up the additional heat from the welding torch, and the hot fluid moves downwards and subsequently strikes the root of the weld, resulting in an enhanced depth of penetration. The fluid flow in the weld pool significantly affects the weld penetration and such a reverse Marangoni flow will result in the weld bead having higher DWR [Lu et al., 2004, 2008, Xu et al., 2007, Kou et al., 1985, Kou and Wang, 1986].

The mere presence of SAE may not be sufficient for the mentioned advantage as a sufficient concentration of SAE is required at the weld pool to reverse the Marangoni currents. Most of the fluxes that are adopted for ATIG/FBTIG/FZTIG are oxides and the minimum concentration of oxygen needed to reverse the Marangoni flow must be understood. An oxygen content between 70 ppm and 200 ppm reverses the temperature coefficient of surface tension (from negative to positive) at the weld pool [Heiple and Roper, 1981; Rückert et al., 2007]. When sulphur-containing compounds are considered as an activating flux, the range of sulphur elements required to cause reverse Marangoni flow is between 5 ppm and 100 ppm [Paillard and Saindrenan, 2003]. In the case of the FBTIG process, the flux gap should be maintained to such an extent that the flux melts/dissociates under the intense heat of the arc and thereby a minimum flux gap becomes necessary to reap the advantages of flux-assisted welding [Santhana Babu et al., 2014]. It is also observed that the quantity of the oxide flux has a significant effect on the penetration depth [Lu et al., 2004]. For flux materials like Cu_2O, NiO and Cr_2O_3, the DWR shows an initial increasing trend which then reduces with the increasing flux quantity. The presence of oxygen in the weld pool was observed to be influential towards the depth of penetration as per the studies. Neither too much nor too little oxygen would enhance the DWR.

Apart from the mentioned factors, buoyancy and electromagnetic forces were also found to weakly affect the depth of penetration. Computational studies have proven that the molten metal flow in the weld pool was affected by electromagnetic forces in addition to the Marangoni convection currents. The weld penetration in alloys of iron and austenitic stainless steels (304 and 316) was found to be controlled by the Marangoni and the buoyancy-driven flow. SAEs such as sulphur and oxygen have a considerable effect on the weld penetration in stainless steels [Pollard, 1988]. Hence, surface tension (γ) can be understood as a function of temperature (T) and the concentration of SAE (C).

4.2.2.1 Insights to Reverse Marangoni Flow

The linear force exerted by a fluid element on its adjacent counterparts results in a reduction of surface area, which is popularly referred to as surface tension. The surface molecules will be pulled down by the immediate subsurface layer, resulting in unbalanced forces of attraction at the exposed surface. However, a molecule deep inside the fluid will be attracted or pulled by adjacent units resulting in an equilibrium state, while the unbalanced forces at the surface cause the fluid to behave like an elastic membrane. This non-equilibrium state causes the surface fluid film to have expendable energy called surface energy. Also, a liquid drop takes the shape of a sphere, which minimizes the surface area and thereby its surface energy.

From the above discussions, it observed that surface tension (γ) is a function of temperature and concentration, i.e., the chemical species present at the surface in reported quantities. The gradient of surface tension will dictate whether the flow will be outward or inward, i.e., Marangoni flow or reverse Marangoni flow. Any variation in the composition along the surface of the weld pool can alter the values of surface tension and can result in the surface tension gradients. Such surface tension gradients caused by the varying amounts of SAE on the weld pool can also initiate fluid flow. Therefore, the surface tension is a function of temperature as well as the composition of the fluid, i.e., $\gamma = f$ (T, C). Here the term C denotes the concentration (quantity) of SAE in the weld pool.

The temperature and the concentration gradient along the weld pool are represented as dT/dS and dC/dS, respectively. The variation of composition along the surface dC/dS can be represented as $(dC/dT) \times (dT/dS)$. The activating flux dissociates due to the intense heat from the welding arc and has a direct relationship with the existing temperature of the weld pool. The SAE released from the dissociating flux is adsorbed on the weld-pool surface, altering the surface tensions. Hence, the relationship of the temperature and concentration gradient at the weld pool determines the direction of fluid flow in the weld pool. The mathematical formulation of the surface tension gradient can be described as in equation (4.1) [Limmaneevichitr and Kou, 2000]:

$$\frac{d\gamma}{dS} = \left(\frac{d\gamma}{dT}\right)_C \left(\frac{dT}{dS}\right) + \left(\frac{d\gamma}{dC}\right)_T \left(\frac{dC}{dS}\right) \qquad (4.1)$$

where the first term depicts the mathematical formulation of the welding process without any surface-active elements or variation in the chemical species along the surface of the weld pool. A negative temperature gradient exists along the spatial distance from the centre line of the weld and therefore the surface tension increases with the spatial distance S. In the absence of any flux containing SAE, the relationship between surface tension and temperature becomes directly proportional. Hence, the flow becomes a Marangoni flow, where the surface tension decreases with the temperature. The studies of Heiple and Roper (1982, 1983) revealed that alumina particles sprinkled over the weld pool were found to move outward from the centre of the weld pool and subsequently sunk into the weld pool as depicted in Figure 4.4e. Such an outward flow results in a wide and shallow weld pool.

In the presence of activating fluxes, the dissociating flux releases SAE into the weld pool, which is subsequently adsorbed on the surface. Not only oxygen, but most of the commonly used SAEs such as sulphur, selenium and tellurium, tend to be adsorbed on the surface of the weld pool. Due to the shear forces of the arc, the distribution of the adsorbed SAE varies along the spatial distance as shown in Figure 4.6. The obvious consequence of this is the reduction of the surface tension forces, but not of the same magnitude as along the surface. The variation in the surface tension forces depends on the effectiveness of adsorption of SAE along the surface. Also, this reinforces the above claim of higher surface tension in the absence of SAE.

The bombardment of charged species on the weld pool imparts arc shear forces on the surface. The momentum of the charged species at the centre of the arc column is high and it manifests

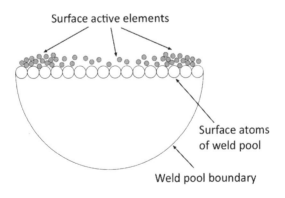

FIGURE 4.6   Variation of SAE along the surface of the weld pool.

itself with higher temperature and also creates arc shear forces. The forces act outwards from the centre, resulting in more accumulation of SAE towards the periphery [Jayakrishnan and Chakravarthy, 2017]. These are macroscopic quantities which possess magnitudes directly proportional to the weld-pool temperature and push the floating SAE from the centre to the periphery of the weld pool as shown in Figure 4.6. This causes a decrease of concentration of SAE which in turn results in increased surface tension at the middle of the weld pool. Reduced segregation of SAE in the weld centre line increases the surface tension when compared to the periphery. On the other hand, higher temperatures along the weld centre line cause the SAE to evaporate at a higher rate resulting in a reduced concentration of SAE at the weld centre line when compared to the periphery (Figure 4.6). Being a complex process, quantitative estimation of the concentration gradients and the evaporation of SAE has not been attempted by other researchers. Due to the presence of SAEs, the surface tension values are higher at the centre of the pool and lower towards the periphery, thereby making the flow inward. On reaching the centre, the molten metal moves downwards deep into the weld pool, causing effective heat transfer to the root of the weld, resulting in increased depth of penetration. Equation (4.2) represents the variations of surface tension gradient with respect to distance from the centre of the pool. While SAEs are present, the concentration gradient terms become significant as they influence the direction of fluid flow. The relationship between concentration gradient, distance and temperature is expressed as

$$\frac{dC}{dS} = \left(\frac{dC}{dT}\right)\left(\frac{dT}{dS}\right) \tag{4.2}$$

The model can be expressed with the governing equation of surface tension-driven fluid flow as

$$\frac{d\gamma}{dS} = \left(\frac{d\gamma}{dT}\right)_c\left(\frac{dT}{dS}\right) + \left(\frac{d\gamma}{dC}\right)\left(\frac{dC}{dT}\right)\left(\frac{dT}{dS}\right) \tag{4.3}$$

The variation of surface tension with the concentration of SAE, variation of concentration of SAE with temperature and the variation of temperature with spatial distance are depicted in Figure 4.7.

As established, the term $(d\gamma/dC)\times(dC/dT)$ represents the temperature dependence of surface tension, which modifies equation (4.3) as

$$\frac{d\gamma}{dS}=\left[\left(\frac{d\gamma}{dT}\right)\left(\frac{dT}{dS}\right)\right]_{\text{without SAE}}+\left[\left(\frac{d\gamma}{dT}\right)\left(\frac{dT}{dS}\right)\right]_{\text{with SAE}} \quad (4.4)$$

Equation (4.4) shows the effect of SAE, which is expressed as two parts, with and without the addition of SAE. The first term of Equation (4.4) without SAE has a positive slope (Figure 4.4d), which causes an outward flow, while the second term with SAE has a negative slope resulting in an inward flow as shown in Figure 4.7.

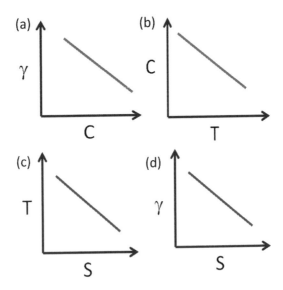

FIGURE 4.7 Variation of surface tension with (a) concentration of SAE, (d) spatial distance, (b) variation of composition with temperature and (c) variation of temperature with spatial distance.

The relative magnitude of these two terms determines the direction of the flow within the weld pool. When the second term (with SAE) dominates, reverse Marangoni flow is set up, which happens when SAE concentration is beyond a threshold value.

4.2.3 Arc Root Constriction

The activating flux is dissociated by the intense heat of the arc and forms a flux cloud enveloping the arc [Vervisch et al., 1990]. The electronegative flux (positive ions) cloud captures the relatively weaker electrons from the periphery of the arc column resulting in narrowing the arc as shown in Figure 4.8.

Electron absorption happens when the positive ions from the cloud interact with the weaker periphery electrons. This phenomenon restricts the current flow to the central regions only, thereby increasing the current density in the plasma. Arc constriction will be more pronounced if the dissociating flux contains molecules or atoms have a large electron attachment cross-section. Thus, halide compounds that have higher electronegativity will have a strong affinity for electrons. Other compounds such as metal oxides, which are not strongly electronegative, but have a higher dissociation temperature, can also be effective in constricting the arc as they can provide a greater number of vaporized ions in the outer regions of the arc. This phenomenon results in reducing the arc diameter and leads to higher current densities [Howse and Lucas, 2000; Simonik, 1976; Fan et al., 2001]. On the other hand, the negatively charged species in the flux cloud also repel the electrons, constricting the arc to narrow diameters. Such arc constriction can invariably result in larger current densities producing welds of large DWR.

4.3 FLUX DEPOSITION

FBTIG welding is based on the concept of selective deposition of flux on the edges to be welded in such a way that a thin narrow strip of the base metal called the flux gap is exposed to the welding arc. Flux gap is an important process parameter that influences

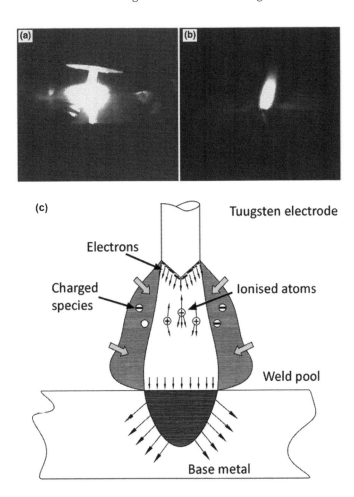

FIGURE 4.8   (a) Profile of the arc during conventional TIG welding, (b) profile of the arc during FBTIG welding ((a) and (b) reprinted from Jayakrishnan, S. et al., *Transactions of Indian Institute of Metals* (2017) 70: 1332, with permission from Springer), (c) Schematic of arc constriction effect.

the depth of penetration and also delineates the process from a normal TIG welding process and ATIG/FZTIG welding process. In general, the fluxes that are used for achieving a larger depth of penetration are metal oxides, metal nitrides and sulphur- and chlorine-containing compounds. These fluxes are available in powder form and are to be selectively applied near the edges to be welded. Neither strong adherence nor a weak adherence of the flux to the base metal is preferred. Strong adherence of the flux to the base metal will hinder easy dissociation of the flux and may result in the entrapment of these particles in the weldment, leading to the deterioration of the weld properties. A very weak adherence of the flux particles will make them get blown away under the influence of arc shear forces and/or due to the flow of shielding gas. To devise an easy methodology, the activating flux is blended properly with a volatile solvent (preferably acetone, isopropyl alcohol or ethanol) to form a paste/slurry and is applied over the base metal using a brush (adopted for lab-scale works mostly). One of the prime properties to be considered in selecting a flux is its vapour pressure, viscosity, thermal conductivity and dielectric constant. Lower vapour pressure and viscosity dictate the effectiveness of flux particles to be bound together and also the effectiveness to retain the same. Thermal conductivity and dielectric constant affect the weld bead properties such as hardness and strength since the flux can determine the way of cooling of the adjacent base metal in the heat-affected zone, thus altering the microstructure. A combination of solvents can also be used and it is observed from a study that a mixture of 40% isopropyl alcohol with acetone by weight provides maximum DWR for nickel-based alloys [Sadewo et al., 2019]. However, there are studies on using flux-water suspension slurry as well as methyl ethyl ketone-flux slurry to coat it on the surface of the base metal [Huang et al., 2012]. Sodium silicate (water glass) was also successfully used as a binder to make a dough out of flux for FBTIG welding [Singh et al., 2017].

A narrow flux gap can be maintained by selectively masking the edges using an adhesive tape and then brushing/spraying the flux over the edges. Later, the adhesive tape is removed to establish the flux gap. The width of the adhesive tape should be carefully chosen to maintain the desired flux gap. Flux gaps ranging from 1 mm to 8 mm in steps of 1 mm can be achieved using this technique. Industrial applications demand methods of application such as spraying using aerosol sprays for uniformity of coating. The doctor blade technique or tape casting technique can also be employed for industrial applications.

4.4 FORCES ACTING ON THE WELD POOL

Four major forces act on the weld pool, namely, Marangoni convection currents due to the existence of surface tension gradient, electromagnetic forces due to the flow of charged species, buoyancy forces which occur due to the density differences in the weld pool and aerodynamic drag forces. A detailed understanding of these forces is necessary to appreciate the performance of flux-activated TIG welding.

4.4.1 Marangoni Convection Currents

The variation of surface tension values across the weld pool governs the direction of Marangoni convection currents. It is established that the surface tension is found to decrease outward in case of TIG welding, causing the flow to be induced radially outwards. Hence, maximum heat transfer is assured towards the walls of the pool leading to wider weld beads rather than deeper ones. This results in low DWR, which is not advisable for any weld joints. Addition of activating fluxes causes a reversal of the flow by changing the direction of surface tension gradient and resulting in higher DWR.

4.4.2 Lorentz/Electromagnetic Forces

During welding, the flow of electrons from the welding torch to the weld material causes an electric field and a concomitant magnetic field. The interaction between these two fields gives rise to electromagnetic forces during welding. Before detachment from the electrode, the molten tip will have direct Lorentz forces within, which can be obtained by integrating the Maxwell stress over its surface. These forces will induce the flow of the gas and arc plasma about the tip of the fire which, in turn, causes indirect electromagnetic forces. If the current density is uniformly distributed over the surfaces, Lorentz force can be expressed as the product of a constant, the square of the total current passing through the tip, and a coefficient which depends upon the shape of the tip. The electromagnetic forces can be a strong factor influencing the resulting weld-pool geometry, depending upon the welding conditions.

4.4.3 Buoyancy Forces

The density difference between various parts of the weld pool causes the buoyancy forces. The upward force caused by lighter fluid and the settling of the heavier part of the fluid is the cause of this buoyancy effect in the weld pool. The centre line of the weld pool receives larger heat input when compared to its periphery and thereby molten metal at the periphery has a higher density than the molten metal at the weld centre. Due to the gravitational forces, the molten metal at the periphery (higher density) sinks to the root of the weld and rises along the weld-pool axis. The density of molten metal decreases with increase in temperature and this fluid tends to flow towards low density. Thus the molten metal moves from the low-temperature region to a high-temperature area of the weld pool based on the density [Oreper et al., 1983].

4.4.4 Aerodynamic Drag

The stream of charged species striking the base metal at the centre creates a force (drag force) which is in a radially outward direction as shown in Figure 4.9. The flow of the shielding gas from the annular region of the torch also imparts shear forces which will be acting in the radially outward direction. This is one of the phenomena in which the hot molten metal from the weld centreline moves towards the periphery causing the weld boundary in the width direction to increase. Similarly, in the flux-assisted welding processes, the distribution of SAE is not found to be of uniform concentration over the surface. The surface-active elements are

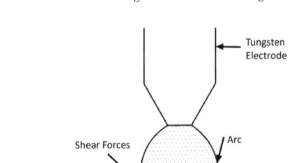

FIGURE 4.9   Direction of aerodynamic forces in a weld pool.

driven to the periphery of the weld due to the arc shear forces and larger adsorption is witnessed at the periphery when compared to the centre line of the weld. The difference in the concentration of SAE induces a surface tension gradient which makes the fluid flow inwards and hence the reverse Marangoni flow.

4.5 FBTIG WELDING PARAMETERS

Many significant factors influence flux-based TIG weld joints, such as the welding current, speed of weld torch movement, flux gap, the particle size of flux materials and coating thickness or density. These parameters are discussed in detail in the following sections.

4.5.1 Welding Current

It is emphasized that the welding current is a significant parameter that decides the heat input to the weld pool. The heat input to the base metal determines the weld bead profile and the microstructure which in turn dictates the performance of the welded structure. The homogeneity of the weld pool/HAZ in terms of composition/microstructure is also affected by the heat that is input to the material. Larger values of depth of penetration and width of the weld pool are observed for higher values of current. The values of depth of penetration and DWR increase for flux-assisted welding processes with an increase in the current values as shown in Figure 4.10 [Tseng, 2013]. The heat that is provided to the weld pool depends on the values of voltage (V) and current (I) during the welding process. Not all the electrical energy is converted to heat. The impedance losses in the circuit and

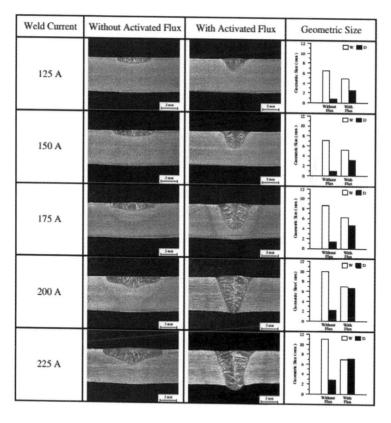

FIGURE 4.10   Effect of welding current on the depth of penetration (D) and width (W) of the weld bead (reprinted from Tseng, K.H., *Powder Technology* (2013) 233: 76, with permission from Elsevier).

the associated energy conversion losses (light and sound) are accounted for using a constant in the corresponding mathematical equation.

The heat flux per unit area, H, is given as:

$$H = \frac{E \times VI}{\pi r^2} \tag{4.5}$$

where E is the efficiency of the welding arc and r is the radius of the arc root. Hence, the heat flux is found to be directly proportional to the applied current in the welding process.

A non-linear Gaussian pattern is observed on the temperature distribution profile in a welding process, which justifies the significantly high temperature rise in the central column compared to peripheral regions of the welding arc. The non-uniform distribution of temperature sets up a temperature gradient in the weld pool and this temperature gradient increases with an increase in the current. In the absence of any SAE, the width of the weld pool increases with a slight increase in the depth of penetration as a result of Marangoni flow. In the presence of SAE, the depth of penetration increases as a result of reverse Marangoni flow [Tanaka et al., 2000]. This relationship of welding current to penetration depth is captured for welding with and without flux and is depicted in Figure 4.10. However, the width of welds achieved for larger currents also seems to be higher. The increase in the width of the molten pool may incorporate the flux elements into the central region of the pool, which increases the electrical impedance of the weld pool. Such an increase in the electrical impedance causes the arc voltage to increase.

4.5.2 Welding Speed

The distance covered by the welding torch per unit of time duration is termed welding speed. Slow traverse of the welding torch increases the heat input to the joint by providing enough time to absorb heat, and hence the welding speed is inversely proportional to the heat input. On the other hand, a higher welding speed reduces the heat input per unit length and hence the depth of penetration decreases. Welding speed becomes an important factor because of the fact that the welding speed controls the amount of heat input to a particular region of the base plate. Since the welding speed alters the heat input to the base metal, the Marangoni forces and the electromagnetic forces in the weld pool are also affected, thereby resulting in a different weld bead profile. It is thus obvious that the shallow weld bead is a result of faster weld torch movement [Vidyarthy and Dwivedi, 2016].

4.5.3 Flux Gap

Flux gap is one of the important parameters/variables in the FBTIG welding process which has a significant effect on the DWR of the weld bead. The larger the flux gap, the lower the DWR; the smaller the flux gap, the larger the DWR. The presence of flux gap also stabilizes the arc on the exposed metal surface. Since the geometry of the weld bead is a combined effect of all three mechanisms in FBTIG welding process, the smaller the flux gap, the larger the insulation effect and arc constriction effect, leading to a larger depth of penetration. The weld bead geometry is also influenced by the nature of flux used and their interaction with the weld pool. It is found that the depth of penetration follows an increasing trend with increasing flux gaps starting from zero and then decreases with further increasing flux gaps. An optimum flux gap for the highest possible penetration can be determined and it depends on the nature of the flux as well as the nature of the base material (Figure 4.11). Studies carried out by Neethu et al. (2019) indicate that the application of silica as the flux with a flux gap of 2 mm yields the highest DWR, whereas the effect of other fluxes (TiO_2 and CaF_2) on DWR is not as pronounced. At all other flux gaps too, silica outperforms other fluxes due to its very low stability as compared to other fluxes. The DWR decreases continuously with an increase in flux gap for both silica and titania as the effect of all mechanisms decreases as the flux

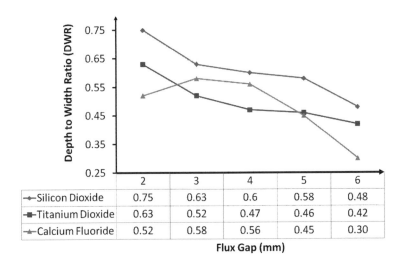

	2	3	4	5	6
─◆─Silicon Dioxide	0.75	0.63	0.6	0.58	0.48
─■─Titanium Dioxide	0.63	0.52	0.47	0.46	0.42
─▲─Calcium Fluoride	0.52	0.58	0.56	0.45	0.30

Flux Gap (mm)

FIGURE 4.11 Effect of flux gap on the DWR during FBTIG welding of AA6061 (reprinted from Neethu, N., Togita, R.G., Neelima, P. et al. *Transactions of Indian Institute of Metals* (2019) 72: 1587, with permission from Springer).

gap increases. However, an anomaly is found in the behaviour of CaF_2 as the flux in which the DWR is higher at 3 mm than 2 mm, after which it decreases. At low flux gaps, there may be a higher concentration of flux particles in the weld pool than in the flux cloud due to the high intensity of the welding arc. But in fluorides, the arc constriction effect is more dominant than the reversal of Marangoni convection currents. This results in a lower DWR at 2 mm. At 3 mm, the current density of the welding arc decreases and hence, the concentration of flux particles in the weld pool and the flux cloud may adjust so that the DWR increases. Following this, DWR decreases due to the overall decrease in the effect of all mechanisms. Such behaviour is not seen in oxide fluxes as for them, the reversal of Marangoni convection currents is a more dominant mechanism than the arc constriction effect.

A flux cloud is formed in the vicinity of the arc column due to the dissociation of flux particles because of the extreme heat from the arc. The arc is constricted by the interaction of this cloud with the low-energy peripheral electrons in the arc, as depicted in Figure 4.8c [Howse and Lucas, 2000; Simonik, 1976, Fan et al., 2001]. The reduction in the flux gap helps more flux particles to participate in the dissociation process and thus leads to a higher concentration of SAE. As the critical concentration of SAE is achieved, the Marangoni flow reversal is observed and results in an increased depth of penetration. The arc root constriction causes an increase in energy density at the weld pool, causing smaller weld bead. Apart from the arc constriction effect, the insulation effect also contributes to a larger depth of penetration with decreasing flux gaps.

4.5.4 Particle Size of Flux Powder

Studies carried out on welding aluminium with silica as the activating flux revealed that the DWR increased with decreasing flux particle size. The reduction in the powder particle size leads to a larger surface area to volume ratio. The larger surface area associated with finer particles leads to higher surface energy and makes the surface atoms having dangling bonds to easily react/dissociate to attain the least possible energy state. Therefore, as the particle attains a minimum size in the order of a few nanometres, the melting point or dissociating temperature decreases. The effective surface area of powder increases as the particle size decreases, which facilitates the easy thermal decomposition of the flux material. This phenomenon is widely represented as melting point depression [Sattler, 2011; Lai et al., 1996]. Finer flux particles dissociate at a much lower temperature and cause an increased concentration of SAE when compared to flux particles of relatively higher size for the same set of welding conditions. Also, the finer flux particles can lead to more electron absorption because of the ease of formation of flux cloud. The combined effect of the above two phenomena enhances the depth of penetration for flux having fine particle size. Figure 4.12 represents the effect of silica particle size on the depth-to-width ratio for aluminium welds. The DWR approaches 1 for a particle size of 600 nm for a flux gap of 3 mm.

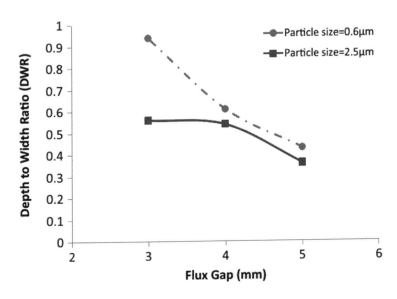

FIGURE 4.12 Influence of flux gap on depth-to-width ratio, compared for two particle sizes (reprinted from Jayakrishnan, S. et al., *Transactions of Indian Institute of Metals* (2017) 70: 1334, with permission from Springer).

4.5.5 Coating Thickness or Density

The amount (mass) of activating flux coated over a unit area of the base metal corresponds to coating density and can be otherwise represented in terms of coating thickness. The intense heat from the arc dissociates the flux particles and releases the SAE in the weld. This means that the thicker layers of the flux coating can contribute towards a higher concentration of SAE in the pool. It is observed that the thickness of flux has a significant influence on the penetration depth. Therefore, there is a minimum quantity of SAE necessary in the weld pool to make the FBTIG mechanisms (reverse Marangoni flow, arc constriction effect and insulation effect) operational, leading to an enhanced depth-to-width ratio [Tanaka et al., 2000]. As the coating thickness increases, the availability of SAE in the weld pool will be higher, which is expected to result in a higher depth-to-width ratio. However, it is observed

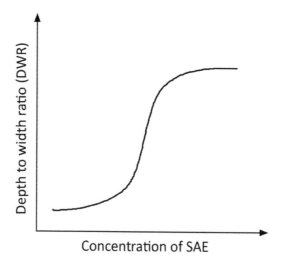

FIGURE 4.13   Depth-to-width ratio of the weld bead for varying coating density.

that an optimal flux thickness exists for maximum depth of penetration, beyond which no significant effect is observed, as depicted in Figure 4.13 for the experiments carried out using TiO_2 as flux. Once the required quantity of SAE is adsorbed over the weld pool, an increase would not result in higher DWR and therefore the curve saturates for a higher quantity of available SAE. Tseng (2013) observed that for a coating density of 0.92–1.86 mg/cm², at weld current values from 125 A to 225 A, the maximum depth-to-width ratio was observed for stainless steels in ATIG welding.

In conjunction with the discussion on welding current, the optimal coating density showed an increasing trend with increasing current. The current required to break through the flux was found to be determined by the coating density, as in case of ATIG welding process. For higher thickness beyond the estimated optimal value, the heat input to the base metal will be affected, resulting in the requirement of higher voltages for the same current to break through the insulating flux of higher coating thickness. This can result in erratic welds due to the arc wandering and extinguishing phenomenon. However, the coating thickness is recommended not to exceed 200 μm. On the other hand, a high coating thickness would also result in the poor appearance of the welds.

4.5.6 Activating Flux

The main requirements for a suitable flux are that its melting temperature should be comparable to that of the melting temperature of the base metal, it should not adversely react with the base metal to form brittle compounds and also should not spoil the weld bead aesthetics. Other required features for fluxes is that they should be electrically insulating and should possess good thermal resistance. The application of activating flux over the weld plates distinguishes the process from a normal TIG welding. Fluxes can drastically alter the weld bead formation by modifying the surface tension values of the molten metal and constricting the arc. The application of flux enhances the weld joint performance by ensuring less heat input to the weld plates, thereby reducing the attendant demerits associated with it and enhancing the weld penetration. Selection of the right flux is a significant step in this process as the DWR is dependent on the type of flux used. Many studies have been carried out in this area, with activating fluxes such as SiO_2, TiO_2, CaO, Al_2O_3, Cr_3O_2, CaF_2, AlF_3, Cu_2O, MgF_2, MoO_3, MoS_2, MnO_2, ZnO, etc., to understand the effect of each flux on the penetration depth. To date, oxides have predominantly been used as the preferred flux. Fluxes are in general non-conducting in their solid state and become conducting in the liquid state. Figure 4.14 depicts the weld bead geometry for metal oxide fluxes such as Cu_2O, NiO, Cr_2O_3, SiO_2 and TiO_2 [Lu et al., 2002].

The depth of penetration initially increases and then decreases with increase in the concentration of oxygen content for most of the fluxes except TiO_2 during welding SS304 plates. However, such a trend is purely dependent on the choice of flux and the base material being welded. The relative stability of the fluxes when exposed to the welding arc can also dictate the amount of SAE release to the weld pool for specific weld parameters. A

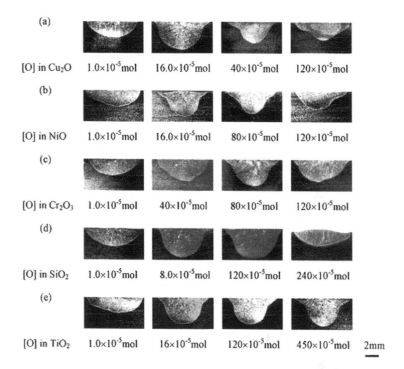

(a)

[O] in Cu₂O   1.0×10⁻⁵mol   16.0×10⁻⁵mol   40×10⁻⁵mol   120×10⁻⁵mol

(b)

[O] in NiO   1.0×10⁻⁵mol   16.0×10⁻⁵mol   80×10⁻⁵mol   120×10⁻⁵mol

(c)

[O] in Cr₂O₃   1.0×10⁻⁵mol   40×10⁻⁵mol   80×10⁻⁵mol   120×10⁻⁵mol

(d)

[O] in SiO₂   1.0×10⁻⁵mol   8.0×10⁻⁵mol   120×10⁻⁵mol   240×10⁻⁵mol

(e)

[O] in TiO₂   1.0×10⁻⁵mol   16×10⁻⁵mol   120×10⁻⁵mol   450×10⁻⁵mol   2mm

FIGURE 4.14   Weld bead geometry study on Cu_2O, NiO, Cr_2O_3, SiO_2 and TiO_2 activated fluxes (reprinted from Lu, S. et al., *Materials Transactions* (2002) 43: 2926, with permission from Japan Institute of Metals).

combination of activating fluxes are also used in varying ratios (for example 1:1 of SiO_2 and MoO_3), apart from the usage of single fluxes. Irrespective of whether the flux is a single-component or a multi-component flux, the presence of fluxes has contributed towards increased depth-to-width ratio. Studies of FBTIG welding on Inconel 718 specimen by Lin et al., 2014 observed that a significant increase in the DWR is possible with a single- or multi-component flux. Use of chlorides as the flux was also found to increase the DWR in the studies of Marya et al. (2002). However, the properties of such welds were specific to various chlorides.

The mechanical properties and weld bead geometry were studied by Huang (2010) and Chern et al. (2011), who found that the

FBTIG welding process is superior to the normal TIG welding process within the range of selected parameters. Chern et al. (2011) studied the effect of SiO_2, MoO_3 and Cr_2O_3 fluxes on 2205 stainless steel welds and found an increase in the penetration depth in combination with improved mechanical strength. Apart from these parameters, the effects of differential heating of the weld surfaces on the amount of angular distortion in the weld plate have been studied. The temperature distribution of the weld pool is non-uniform as the exposed area (unmasked by flux) gets cooled faster than the inner core, by convection/radiation. This differential heat causes shrinkage of top layers whereas the inner layers remain hotter, resulting in the development of thermal stresses from the temperature gradients. The thermal stress must be limited well below the yield strength of the base metal to avoid warpage/distortion [Mollicone et al., 2006]. The presence of tensile residual stresses on the surface beyond a threshold magnitude can also lead to premature failures during service. The presence of activating fluxes on the surface facilitates the reduction of angular distortion and development of residual stresses as compared to conventional TIG welding [Huang, 2010]. However, the extent of decrease depends on the nature of flux and its thermal conductivity in the right combination with the base metal.

Another study by Tseng and Hsu (2011) also reinforces the above concept that the use of activating fluxes increases the weld penetration and decreases the angular distortion. The Ellingham diagram (Figure 4.15) that defines the relative stability of the oxides is used as a tool to select the right of flux based on the ease of decomposition [Seetharaman, 2014]. Since the depth of penetration in flux-assisted welding depends on the optimum amount of SAEs present in the weld pool, their decomposition at the said weld conditions needs to be ascertained. The ease of decomposition of each flux, shown in the Ellingham diagram, assists in choosing the flux. Lu et al. (2003) report the relationship between the particle size of the flux and its stability. The study revealed that the ease of decomposition of fluxes increases with

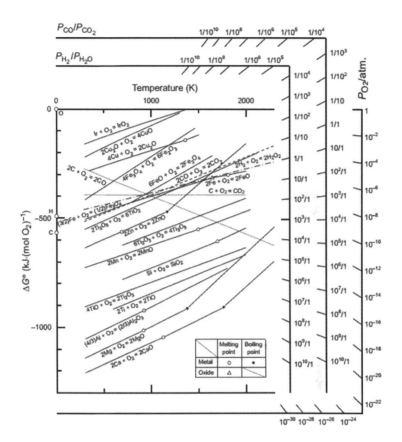

FIGURE 4.15 Ellingham diagram (reprinted from Hasegawa, M., *Treatise on process metallurgy volume 1: process fundamentals* (2014), 508, with permission from Elsevier).

a reduction in particle size. It can be observed that flux of finer particle size allows easy decomposition under the intense heat of the arc resulting in the prominent effect on flux-assisted welding processes. This can be attributed to the fact that finer particles have a larger surface area and therefore their stability decreases, leading to easy decomposition.

The less stable metal oxides occupy higher positions in the Ellingham diagram as shown in Figure 4.15. Silica, calcium oxide and titania are the most commonly used flux materials, out of which silica has the lowest bond dissociation energy (less stable), which makes it a strong candidate for the choice of flux in flux-assisted welding processes. Since it has low dissociation energy, silica decomposes with lesser energy input and also ensures higher concentrations of SAE at the weld pool. This facilitates a lower amount (coating thickness) of flux at the joint surface for the FBTIG welding process, which provides the required SAE. Thinner coating also ensures less entrapment of flux particles in the case of ATIG welding. When the oxides that occupy lower positions in the Ellingham diagram are used as fluxes, thicker coatings are required with higher current and voltages during the welding process to ensure the critical amount of SAE. Neethu et al. (2019) investigated the shape of the welding arc for different fluxes as depicted in Figure 4.16. The arc formed during TIG welding is much wider than those using fluxes. When an activating flux is used, the arc constricts and caves causing an increased

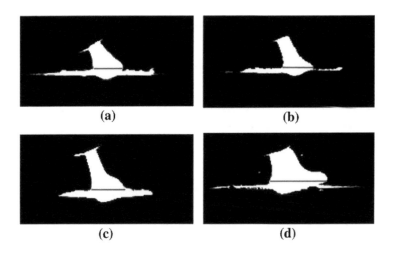

FIGURE 4.16 Images of welding arc profiles during FBTIG welding for a flux gap of 3 mm for (a) SiO_2 flux, (b) CaF_2 flux, (c) TiO_2 flux, (d) no flux (reprinted from Neethu, N., Togita, R.G., Neelima, P. et al. *Transactions of Indian Institute of Metals* (2019) 72: 1587, with permission from Springer).

DWR. This happens because of the influence of insulation effect and the arc constriction effect. The former channels the arc root to the flux gap while the latter constricts the overall distribution of the arc. The most constricted arc is observed for SiO_2 which yields a higher DWR. The low stability of SiO_2 paves way for easier decomposition and better consequences of the mechanisms.

Yong et al. (2007) has developed a multi-component flux, AF305 (composition not revealed), and observed that a larger depth of penetration is possible when compared to the use of SiO_2 as a flux for welding aluminium alloys. The use of fluorides as flux revealed that the activating effect is mainly because of their effects on the arc physics. Since fluorides are strongly electronegative, arc constriction is predominant in bringing about a change in the depth of penetration. Thus, the cost and complexities of the welding process are also associated with the choice of flux materials. Therefore, it is important to select the right choice flux to have a maximum depth of penetration or higher DWR. Development of newer multi-component fluxes is also a possible step to increase the depth of penetration for particular materials.

4.6 PROS AND CONS OF FBTIG OVER TIG/ATIG/FZTIG

The depth of penetration and DWR are observed to be higher in the case of flux-assisted welding process when compared to a normal TIG welding process. The presence of flux facilitates welding of thicker metal sections efficiently with lower heat input than the normal TIG welding process and this also results in distortion-free weldments. The application of flux makes the mechanisms such as reverse Marangoni convection, arc constriction and insulation effect operational. While all three mechanisms are operational for the FBTIG welding process, only the reverse Marangoni convection and arc constriction effects are possible for ATIG/FZTIG. The insulation effect is an additional mechanism for the FBTIG process over ATIG/FZTIG because of the presence of the flux gap, where the arc is channelled to strike the base metal in the narrow flux gap to follow the path of least resistance for the charged species. The presence of insulation effect in FBTIG welding makes this process more effective in terms of depth of penetration and many other advantages compared to the other two. FBTIG welding of Inconel 718 with a 1.2 mm flux gap showed a slender bead geometry compared to that in the ATIG process, as reported by Lin and Wu (2012). The weld bead images observed at different locations of the weld are shown in Figure 4.17 and it is very evident that FBTIG welds have higher DWRs. The clear advantage of the FBTIG welding process is observed from an Inconel 718 alloy plate weld joint which gave complete penetration of 6.35 mm in a single pass.

The process of ATIG welding requires complete masking of the edges to be welded by the flux which obstructs the vision of the operator to traverse the torch along the edges to be joined. The flux coating on the edges in the case of the ATIG/FZTIG process mandates the process to be automated. The expense incurred and the complexities involved make ATIG/FZTIG welding a non-versatile and less appealing process for industries to adopt and practice.

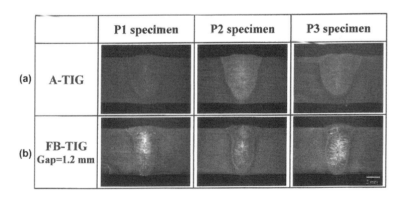

FIGURE 4.17 Comparison of weld beads for (a) ATIG welding, (b) FBTIG welding on INCONEL alloy welds (reprinted from Lin, H.-L., Wu, T.-M., *Materials and Manufacturing Processes* (2012) 27(12): 1460, with permission from Taylor and Francis).

FBTIG welding overcomes this limitation and is a feasible alternative for industries which do not require the flux to be masked on the edges to be welded and thereby provides enough visibility to the operator to weld under close observation. The additional mechanism (insulation effect) in FBTIG welding process is a boon which is inherent in the process by itself. Since the flux bounds the arc, the weld pool will not spread in the width direction for thicker flux coatings and will be bounded by the narrow flux gap. For thicker welds, thicker coatings are required in ATIG/FZTIG, where the higher thickness of the flux coating requires more dissociation energy in the form of heat. Once the flux coating thickness exceeds the critical value, a high value of current is required to dissociate the flux to impart SAE to the weld pool. FBTIG welding provides a solution for such situations by having a thin flux coating which stabilizes the plasma and prevents the extinguishing of the arc during the process. On the other hand, in the case of the ATIG/FBTIG welding process, the arc is struck on the flux coating which requires higher values of current to initiate and sustain the arc for the entire bead formation. The arc follows the least resistant path and tends to wander if the coating thickness is not kept uniform over the surface. No tolerance in the coating thickness is permitted in the case of the ATIG/FZTIG process; otherwise, arc wandering may occur. There is a possibility that arc wandering can make welds with inferior properties. The complete masking of flux in ATIG/FBTIG welding makes the process more vulnerable, wherein the probability of entrapment of flux particles in the weldment is high and thereby worsens the weld properties. Whereas in FBTIG welding process, the arc is struck over the base metal only and the possibility of flux entrapment in the weld bead is very low. The coating thickness window in ATIG welding is much lesser, which implies that strict control of coating thickness is to be maintained. The FBTIG welding process produces comparable results even for larger coating thickness, thereby inferring that the process is less sensitive to coating thickness. This enables a larger window of coating thickness in the FBTIG welding process [Singh et al., 2017]. However, any flux-assisted welding process has its demerits such as the requirement of multistage welding operations (flux coating adds to the process as a primary stage), higher time consumption as compared to normal TIG welding and reduced production efficiency (as the number of products will be less than for the normal welding techniques). The presence of sulphur in the joint increases the arc voltage thereby resulting in higher heat input, giving rise to increased δ ferrite in steels, which is the reason for hot cracking [Snow, 2002].

Yet another difficulty in flux-assisted welding processes is the difficulty in the deposition of flux. Industries may adopt a cold spray technique to enhance productivity. If an automated spraying technique is adopted, it facilitates the ATIG welding process since the entire edges to be welded are to be masked with the flux. In the case of FBTIG welding, a narrow flux gap is to be maintained and a suitable mask is required before spraying to establish the flux gap. Also, strict control of the flux gap along the edges to be welded is required for the FBTIG welding process. All the flux-assisted welding processes are feasible for welding plates of sufficient thickness which do not require any edge preparation. This attributes the applicability of flux-assisted welding process to single-pass welding without any difficulty. For welding thicker plates which require the edges to be prepared, the application of flux on the slanting edges and the root becomes difficult and at the same time, the flux has to be coated on the completion of every single pass, which decreases the productivity. Such flux-assisted welding processes become difficult when welding is performed in the overhead position and also, the process is more friendly upon welding flat plates. Due to the presence of a thermally insulating flux coating over the base metal, the grain size in the vicinity of the heat-affected zone for FBTIG-welded specimens is generally found to be slightly larger than in TIG-welded specimens. However, the improvement of yield strength in a few aluminium alloy welds performed by the FBTIG welding process was attributed to the fine-grained structure. When significance is given to weld aesthetics or weld bead appearance, TIG-welded specimens outweigh all the other mentioned process. The presence of flux always deteriorates the surface appearance because of its decomposition and leaves stain marks on the surface. However, this effect is more pronounced in the case of ATIG welding and much less pronounced in FBTIG welding where the flux is not directly exposed to the arc. However, the welding polarity also plays a significant role in the weld bead aesthetics in the case of flux-assisted welding. Welding with an AC power source has a good combination of cleaning effect with deeper penetration, whereas the direct current straight polarity (DCSP) produces welds having less cleaning effect with deeper penetration and direct current reverse polarity (DCRP) produces welds of optimum quality with shallow penetration. The problem of electrode burn-out and contaminating the weld is also possible with DCRP. Yet another feature of this activated TIG welding process is that the depth of penetration can be varied by altering the composition of the shielding gas. Studies carried out by Huang (2009) on welding austenitic stainless steels revealed that there was a notable change in the depth of penetration when argon along with 10 volume% nitrogen was used as a shielding gas. The increase in the depth of penetration was attributed to larger heat content carried by nitrogen species. The larger heat content also resulted in higher angular distortion welds as compared to conventional TIG welding. When hydrogen is used along with argon, the weld bead width, as well as penetration, was observed to increase.

4.7 FBTIG: GROWTH AND ADVANCEMENTS

Sire et al. (2001) developed this process as a variant of the ATIG process and observed that by having a narrow flux gap, the depth of penetration can be further increased. They initially demonstrated the feasibility of this process on aluminium alloy 5086 and reported that penetrations as deep as 6 mm can be achieved in a single run for a flux gap of 4 mm, which was not possible in the case of a conventional TIG welding process. The FBTIG welding process improves not only the depth of penetration but also the DWR. Hence, it can be generalized that for given conditions, FBTIG welding generates a much higher depth of penetration and DWR as compared to the conventional TIG welding process. The channelling of the welding arc along the flux gap and the increased powder density are attributed to the better DWR in FBTIG welding compared to conventional methods. It is also observed that the FBTIG welding arc is much more stable than that of the arc established in TIG welding. The studies of Zhao et al. (2011) showed that single-component fluxes like SiO_2, TiO_2, CaF_2, MgO and NaCl result in increased depth of penetration for various possible flux gaps. Among the other oxide fluxes that are used, FBTIG welding with silica as a flux has shown the maximum depth of penetration for fixed weld conditions. However, the width of weld varied for different single-component fluxes used and it is purely dependent on the activating mechanisms that are operative for each type of flux. It is also mentioned that the voltage levels required to establish/sustain the arc over the flux covered zone are higher when compared to a normal TIG welding process that does not have a flux cover. The arc constriction effect was also observed when the arc moves over the flux covered zone, resulting in higher current density. The effects of various process parameters of FBTIG welding were studied by Rückert et al. (2014) and it was observed that the flux gap plays a pivotal role in increasing the depth of penetration. The flux gap was varied within a possible range and it was observed that when the flux gap is comparable to the diameter of the welding arc, the arc voltage also varies and results in a higher depth of penetration. The influence of particle size of the flux on weld bead geometry was studied by Jayakrishnan et al. (2016) who found that the finer the particles, the deeper the penetration. This was attributed to the fact that the finer flux particles dissociate at a lower heat input, leading to higher SAE concentration in the weld pool. Sire and Marya (2002) found that the values of DWR and depth of penetration for FBTIG are twice those of the normal TIG welding process due to the high electrical resistivity of the flux coating and the arc confinement in the case of the FBTIG welding process. Studies carried out by Babu et al. (2014) revealed that the tensile properties of FBTIG-welded specimens were found to be superior to those of a TIG-welded specimen. FBTIG-welded specimens for a flux gap of 4 mm have shown a 10% increase yield strength compared to TIG-welded specimens. Finer grains were observed in the weld zone, which attributes for the increased strength for FBTIG-welded specimens. The variation in the coating thickness affects the electrical resistance, thereby resulting in the arc wandering and hence, inferior weld strength in the case of the ATIG welding process, whereas the welds performed by the FBTIG welding process did not exhibit such phenomena. Liquation cracks were observed in the specimen welded by the ATIG process, whereas FBTIG specimens were free from this tendency. Corrosion studies performed on the TIG-welded joints and FBTIG-welded joints revealed a comparable resistance to stress corrosion cracking and no considerable reduction in the ductility was observed for the FBTIG-welded specimens even after exposure to a corrosive medium. Apart from the parameters (flux gap and flux particle size) which are exclusive for FBTIG welding, the peak welding current and welding speed were also found to have a significant impact on the integrity of the joint [Babu et al., 2014]. Arc wandering is a phenomenon that is observed in the case of ATIG welding and predominantly absent in FBTIG welding due to the arc being struck on the exposed base metal in the flux gap. Various modelling methods were developed to connect the process parameters like welding current, welding speed, etc. with the bead geometry.

One of the significant mechanisms for the improvement of depth of penetration in the case of FBTIG welding is the insulation effect and the main reason for achieving an improved bead surface quality is because the flux is spread on either side of the joint and not over the joint as in the case of ATIG welding [Vilarinho et al., 2009]. When the arc is struck on base metals in the narrow flux gap, the weld bead appearance is more likely to resemble that of a normal TIG welding process, wherein the appearance of an ATIG weld bead is relatively shabby.

Some typical applications of flux-assisted processes include nuclear reactor components, car wheel rims, steel bottles, pipelines and pressure vessels as reported by Lucas and Howse (1996). AISI type 316 stainless steel tube of 5 mm wall thickness was reported to be welded in a single pass without using any filler wire by using the activating flux under the conditions of pulse current 150 A, background current 30 A, arc voltage 9.5 V and welding speed 60 mm/min (Howse and Lucas, 2000). However, it is observed that this work was carried out in a laboratory and also on a prototype basis. The commercialization of such processes needs to happen in industry to witness a major change in productivity. Industries having automated welding systems can adopt either the ATIG welding process or the FBTIG welding process. But the FBTIG welding process is proven to provide enhanced penetration because of the additional insulation effect. Commercialization of the FBTIG welding process is much easier as it doesn't require any automation. Further, small-scale industries can reap the benefits of this process.

4.8 NUMERICAL SIMULATIONS OF FBTIG WELDING

Numerical modelling of welding is a complex problem involving heat transfer, fluid flow, mass transfer, phase transfer, etc. Welding is essentially modelled as a three-dimensional transient problem. Finite element models are usually used to predict thermo-mechanical changes in the weld whereas finite volume models are better suited for modelling the fluid melting, flow and solidification in the weld pool.

For modelling the heat transfer within the material, it is necessary to identify the various heat sources and sinks. During welding, heat is input to the material by the welding arc and lost to the surroundings by convection and radiation. It is also transferred within the material by conduction. An important factor that affects the predicted weld bead shape in simulations is the model of the input heat source. The heat source can be modelled as a surface or a volumetric phenomenon with spatial power distribution moving at a given traverse rate. Additionally, there can be temporal changes such as pulsing effects. The amount of heat absorbed into the material is determined by the absorption efficiency, which depends on the material, ambient conditions such as temperature, etc. The effect of shielding gases, electrode tip, etc. can also be taken into consideration for developing an accurate model.

Many mathematical models have been proposed to simulate the welding arc. Among these, an efficient model is the Pavelic disc model in which, for an arc moving on the x–z plane, the surface heat flux, q (W/m²), is given by

$$q(x,z,t) = \frac{3Q}{\pi c^2} e^{-3x^2/c^2} e^{-3[z+v(\tau-t)]^2/c^2}$$

where Q is the energy input rate in watts, c is the characteristic radius of distribution in meters and τ is the lag factor in seconds.

However, it is observed from experiments that the heat distribution function has a better fit if it is assumed to have an ellipsoidal distribution when compared to the disc model. It is also evident that the power density assumes a different value in the front and rear of the welding arc due to its traversing nature. To account for this, the double ellipsoidal model was proposed, which has enjoyed considerable popularity. In this method, the heat source is modelled using front and rear ellipsoids with different power distributions. The surface flux, q (W/m²) for the front quadrant using this model is given by

$$q(x,y,z,t) = \frac{6\sqrt{3}f_f Q}{abc\pi\sqrt{\pi}} e^{-3x^2/a^2} e^{-3y^2/b^2} e^{-3[z+v(\tau-t)]^2/c^2} \qquad (4.6)$$

Similarly for the rear quadrant,

$$q(x,y,z,t) = \frac{6\sqrt{3}f_r Q}{abc\pi\sqrt{\pi}} e^{-3x^2/a^2} e^{-3y^2/b^2} e^{-3[z+v(\tau-t)]^2/c^2} \qquad (4.7)$$

where f_f and f_r are the fractions of heat deposited in the front and rear quadrants, and a, b and c are the semi-axes of the ellipsoids. Q and τ hold the same meaning as in the Pavelic model.

These models fit the welding arc during TIG welding appreciably well. However, during FBTIG welding, arc constriction and insulation effects arise which alter the shape of the welding arc. Instead of gaining the 'bell-shaped' curve commonly observed, the root of the arc is channelled to fit the flux gap to choose the least resistant path. Numerically simulating FBTIG welding will require mathematical quantification of the welding arc so that it can be given as a heat flux boundary condition. However, no theory has yet been proposed for this.

The reverse-engineering approach can be used to model the FBTIG welding arc. An image of the arc can be captured by suitable means during welding to which an appropriate mathematical model can be developed to predict the shape and thus heat flux that is input to the material. This model can then be used for numerical simulation. The weld pool and its dynamics also need to be considered for modelling any type of welding. Major forces that are present in the weld pool are buoyancy, Marangoni convection currents and electromagnetic forces. Apart from these, there are minor forces such as due to gas flow, rigid body motion, etc. It is generally assumed that the fluid flow in a weld pool is of a Newtonian fluid undergoing incompressible laminar motion, driven by the currents within the melt. The motion is governed by the Navier–Stokes equation.

In TIG welding, the gradient of surface tension with temperature $(d\gamma/dT)$ is negative. This causes Marangoni convection currents to flow from the centre of the weld pool to the periphery causing a low depth-to-width ratio (DWR). Due to the presence of activating fluxes, $(d\gamma/dT)$ becomes positive. This causes a reversal in the direction of the Marangoni flow, resulting in an increased DWR.

Figure 4.18a shows a Marangoni flow in an AA6061 weld pool where the gradient of surface tension with temperature is $d\gamma/dT = -0.000128$ N/mK. The motion of the fluid from the centre

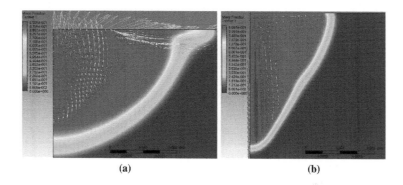

(a)   (b)

FIGURE 4.18 Numerical simulation of (a) Marangoni flow, (b) reverse Marangoni flow (reprinted from Neethu, N., Togita, R.G., Neelima, P. et al. *Transactions of Indian Institute of Metals* (2019) 72: 1588, with permission from Springer).

to the periphery is evident from the velocity vectors. This leads to a shallow weld pool resulting in lower DWR. Figure 4.18b shows a reverse Marangoni flow weld pool with $d\gamma/dT = 0.00064$ N/mK. The reverse flow transfers heat from the arc to the weld root, increasing its depth and decreasing the width. This gives a geometry with a higher DWR and, hence, better weld properties. A complete numerical simulation of FBTIG welding will involve coupled solutions that consider the arc constriction effect, insulation effect and reverse Marangoni flow as well as the dissociation behaviour and motion of the flux particles under the intense heat of the welding arc. Simulations can even be used for prediction of flux entrapments within the weld pool. Modelling all phenomena involved is a complex and interesting problem which has not yet been attempted. An accurate numerical solution will help in the development of FBTIG welding.

4.9 FBTIG: FUTURE OF THE TECHNIQUE

The weld parameters common to FBTIG and ATIG welding processes need to be varied in combinations and need to be explored for understanding the advantages of the processes. Very few studies are available regarding the insulation effect observed in FBTIG welding which is due to the caving in of the arc causing higher energy density at the arc root. Being unique to FBTIG welding, this is a distinguishing factor between ATIG and FBTIG welding. A detailed understanding of the effect of fluxes on the weld bead geometry, which can be beneficial for the improvement of weld strength, is presently absent. A thorough correlation between theoretical and experimental results for FBTIG weld parameters, forces affecting the parameters and the weld geometry is required. The effects of Lorentz forces, surface tension, the combination of fluxes, etc. need to be mathematically modelled and to be validated experimentally. There are very few studies on the numerical simulation of the weld geometry and alteration of the fusion line boundary coupled with the convection currents. Analytical understanding by microscopical observation, inclusion density and studies on microstructure are very significant in the case of flux-assisted welding processes.

References

Antonini JM (2014), *Comprehensive Materials Processing*; Vol 8, Elsevier Publications.

Brungraber RJ, and Nelson FG (1973), Effect of welding variables on aluminum alloy weldments. *Welding Research*; Vol 97, pp 97–103.

Chern Tsann-Shyi, Tseng Kuang-Hung, and Tsaia Hsien-Lung (2011), Study of the characteristics of duplex stainless steel activated tungsten inert gas welds. *Materials and Design*; Vol 32(1), pp 255–263.

den Ouden G, and Hermans MJM (2009), *Welding Technology*, 1st Edition, Delft University of Technology, the Netherlands.

Fan Ding, Zhang Ruihua, Gu Yufen, and Ushio Masao (2001), Effect of flux on A-TIG welding of mild steels. *Transaction of JWRI*; Vol 30(1), pp 35–40.

Heiple CR, and Roper JR (1981), Effect of selenium on GTAW fusion zone geometry. *Welding Journal – Research Supplement*; Vol 60, pp 143s–145s.

Heiple CR, and Roper JR (1982), Mechanism for minor element effect on GTA fusion zone geometry. *Welding Journal – Research Supplement*; Vol 61, pp 97s–102s.

Heiple CR, Roper JR, Stagner RT, and Aden RJ (1983), Surface active element effects on the shape of GTA, laser, and electron beam welds. *Welding Journal*; Vol 62, pp 72s–77s.

Howse DS, and Lucas W (2000), Investigation into arc constriction by active fluxes for tungsten inert gas welding. *Science and Technology of Welding and Joining*; Vol 5(3), pp 189–193.

Huang Her-Yueh (2009), Effects of shielding gas composition and activating flux on GTAW weldments. *Materials and Design*; Vol 30(7), pp 2404–2409.

Huang Her-Yueh (2010), Effects of activating flux on the welded joint characteristics in gas metal arc welding. *Materials and Design*; Vol 31(5), pp 2488–2495.

Huang Y, Fan D, and Shao F (2012), Alternative current flux zoned tungsten inert gas welding process for aluminium alloys. *Science and Technology of Welding and Joining*; Vol. 17(2), pp 122–127.

Jayakrishnan S, and Chakravarthy P (2017), Flux bounded tungsten inert gas welding for enhanced weld performance—A review. *Journal of Manufacturing Processes*; Vol 28, pp 116–130.

Jayakrishnan S, Chakravarthy P, and Rijas AM (2016), Effect of flux gap and particle size on the depth of penetration in FBTIG welding of aluminium. *Transactions of Indian Institute of Metals*; Vol 70(5), pp 1329–1335.

Kou S, and Sun DK (1985), Fluid flow and weld penetration in stationary arc welds. *Metallurgical Transactions A*; Vol 16(1), pp 203–213.

Kou S, and Wang YH (1986), Weld pool convection and its effect. *Welding Journal – Research Supplement*; Vol 65, pp 63s–70s.

Lai SL, Guo JY, Petrova V, Ramanath G, and Allen LH (1996), Size-dependent melting properties of small tin particles: Nanocalorimetric measurements. *Physical Review Letters*; Vol 77(1), pp 99–102.

Leconte S, Paillard P, and Saindrenan J (2013), Effect of fluxes containing oxides on tungsten inert gas welding process. *Science and Technology of Welding and Joining*; Vol 11(1), pp 43–47.

Limmaneevichitr Chaowalit, and Kou Sindo (2000), Visualization of marangoni convection in simulated weld pools containing a surface-active agent. *Welding Journal*; Vol 79(11), pp 324s–330s.

Lin Hsuan-Liang, and Wu Tong-Min (2012), Effects of activating flux on weld bead geometry of inconel 718 alloy TIG welds. *Materials and Manufacturing Processes*; 27(12), pp 1457–1461.

Lin Hsuan-Liang, Wu Tong-Min, and Cheng Ching-Min (2014), Effects of flux precoating and process parameter on welding performance of Inconel 718 alloy TIG welds. *Journal of Materials Engineering and Performance*; Vol 23(1), pp 125–132.

Lowke JJ, Tanaka M, and Ushio M (2004), Insulation effects of flux layer in producing greater weld depth. *The 57th Annual Assembly of International Institute of Welding*, 2004 (Osaka), IIW Doc.212-1053-04, 2004.

Lowke JJ, Tanaka M, and Ushio M (2005), Mechanisms giving increased weld depth due to a flux. *Journal of Physics D: Applied Physics*; Vol 38(18), pp 3438–3445.

Lu Shanping, Fujii Hidetoshi, Sugiyama Hiroyuki, Tanaka Manabu, and Nogi Kiyoshi (2002), Weld penetration and marangoni convection with oxide fluxes in GTA welding. *Materials Transactions*; Vol 43(11), pp 2926–2931.

Lu Shanping, Fujii Hidetoshi, Sugiyama Hiroyuki, and Nogi Kiyoshi (2003), Mechanism and optimization of oxide fluxes for deep penetration in gas tungsten arc welding. *Metallurgical and Materials Transactions A*; Vol 34(9), pp 1901–1907.

Lu Shanping, Fujii Hidetoshi, and Nogi Kiyoshi (2004), Marangoni convection and weld shape variations in Ar-O_2 and Ar-CO_2 shielded GTA welding. *Materials Science and Engineering A*; Vol 380(1), pp 290–297.

Lu Shanping, Fujii Hidetoshi, and Nogi Kiyoshi (2008), Marangoni convection and weld shape variations in He-CO_2 shielded gas tungsten arc welding on SUS304 stainless steel. *Journal of Materials Science*; Vol 43(13), pp 4583–4591.

Lucas W, and Howse D (1996), Activating flux-increasing the performance and productivity of the TIG and plasma processes. *Welding and Metal Fabrication*; Vol 64(1), pp 11–17.

Marya M, and Edwards GR (2002), Chloride contributions in flux-assisted GTA welding of magnesium alloys. *Welding Journal*; Vol 81(12), pp 291s–298s.

Messler Robert W, Jr. (2008), *Principles of Welding: Processes, Physics, Chemistry and Metallurgy*, John Wiley & Sons, Singapore.

Modenesi Paulo J (2015), The chemistry of TIG weld bead formation. *Welding International*; Vol 29(10), pp 771–782.

Modenesi Paulo J, Apolinário Eustáquio R, and Pereira Laci M. (2000), TIG welding with single-component fluxes. *Journal of Materials Processing Technology*; Vol 99(1), pp 260–265.

Mollicone P, Camilleri D, Grayand TGF, and Comlekci T (2006), Simple thermo-elasticplastic models for welding distortion simulation. *Journal of Materials Processing Technology*; Vol 176(1–3), pp 77–86.

Neethu N, Togita Rahul Goud, Neelima P, Chakravarthy P, Narayana Murty SVS, and Nair Manoj T. (2019), Effect of nature of flux and flux gap on the depth-to-width ratio in flux-bounded TIG welding of AA6061: Experiments and numerical simulations. *Transactions of Indian Institute of Metals*; Vol 72(6), pp 1585–1588.

Norrish John (2006), *Advanced Welding Processes*, Woodhead Publishing Limited, Cambridge, UK.

O'Brien Annette (2004), Welding Handbook, *Welding Processes*, Part-1, Vol 2, 9th Edition, American Welding Society, Miami.

Oreper GM, Eagar TW, and Szekely J (1983), Convection in arc weld pools. *Welding Journal – Welding Research Supplement*; Vol 62(11), pp 307–312.

Paillard P, and Saindrenan J (2003), Effect of activating fluxes on the penetration capability of the TIG welding arc: Study of fluid-flow phenomena in weld pools and the energy concentration in the anode spot of a TIG arc plasma. *Materials Science Forum*; Vol 426–432, pp 4087–4092.

Paskell T, et al. (1997), GTAW flux increases weld joint penetration. *Welding Journal*; Vol 76(4), pp 57–62.

Pollard B (1988), The effects of minor elements on the welding characteristics of stainless steel. *Welding Journal*; Vol 67(9), pp 202s–213s.

Rückert G, Huneau B, and Marya S (2007), Optimizing the design of silica coating for productivity gains during the TIG welding of 304L stainless steel. *Materials and Design*; Vol 28(9), pp 2387–2393.

Rückert G, Perry N, Sire S, and Marya S (2014), Enhanced weld penetrations in GTA welding with activating fluxes case studies: Plain carbon & stainless steels, titanium and aluminum. *Materials Science Forum*; Vol 783–786, pp 2804–2809.

Sadewo AP, Ardhyananta H, and Purniawan A (2019), Effect of acetone/isopropanol as solvent TiO_2 active flux on penetration characteristics activated tungsten inert gas welding Incoloy 825. *IOP Conference Series: Materials Science and Engineering*; Vol 546, 042039.

Santhana Babu AV, Giridharan PK, Ramesh Narayanan P, Narayana Murty SVS, and Sharma VMJ (2014), Experimental investigations on tensile strength of flux bounded TIG welds of AA2219-T8 aluminum alloy. *Journal of Advanced Manufacturing Systems*; Vol 13(2), pp 103–112.

Sattler KD (2011), *Handbook of Nanophyics, Nanoparticles and Quantum Dots*, CRC Press, Taylor & Francis Group, Boca Raton, FL.

Seetharaman Seshadri (2014), *Treatise on Powder Metallurgy Vol 1: Process Fundamentals* (Edited by Masakatsu Hasegawa), Elsevier.

Simonik AG (1976), The effect of contraction of the arc discharge upon the introduction of electro-negative elements. *Welding Production*; Vol 23, pp 49–58.

Singh Akhilesh Kumar, Dey Vidyut, and Rai Ram Naresh (2017), A study to enhance the depth of penetration in Grade P91 steel plate using alumina as flux in FBTIG welding. *Arabian Journal for Science and Engineering*; Vol 42(11), pp 4959–4970.

Sire S, Marya S, and Ohji T. (2001), New perspectives in TIG welding of Aliuminium through flux application FBTIG process. Proc. of the 7th Int Symp, 113–118, JWS, Kobe Japan.

Sire Stephane, and Marya Surendar (2002), On the development of a new flux bounded TIG process (FBTIG) to enhance weld penetrations in aluminium 5086. *International Journal of Forming Processes*; Vol 5(1), pp 39–51.

Snow Heather M, (Marie) (2002), Investigation of the effect of a surface active flux on the microstructure and properties of gas tungsten arc welds made on a superaustenitic stainless steel. Master of Science Thesis, Lehigh Unversity.

Tanaka M, Shimizu T, Terasaki T, Ushio M, Koshiishi F, and Yang CL (2000), Effects of activating flux on arc phenomena in gas tungsten arc welding. *Science and Technology of Welding and Joining*; Vol 5(6), pp 397–402.

Tsai MC, and Kou Sindo (1989), Marangoni convection in weld pools with a free surface. *International Journal for Numerical Methods in Fluids*; Vol 9(12), pp 1503–1516.

Tseng Kuang-Hung (2013), Development and application of oxide-based flux powder for tungsten inert gas welding of austenitic stainless steels. *Powder Technology*; Vol 233, pp 72–79.

Tseng Kuang-Hung, and Hsu Chih-Yu (2011), Performance of activated TIG process in austenitic stainless steel welds. *Journal of Materials Processing Technology*; Vol 211(3), pp 503–512.

Vervisch P, Cheron B, and Lhuissier JF (1990), Spectroscopic analysis of a TIG arc plasma. *Journal of Physics D: Applied Physics*; Vol 23(8), pp 1058–1063.

Vidyarthy RS, and Dwivedi DK (2016), Activating flux tungsten inert gas welding for enhanced weld penetration. *Journal of Manufacturing Processes*; Vol 22, pp 211–228.

Vilarinho Louriel O, Raghunathan Sayee, and Lucas Bill (2009), Observation of insulation mechanism during FBTIG. 20th International Congress of Mechanical Engineering.

Xu YL, Dong ZB, Wei YH, and Yang CL (2007), Marangoni convection and weld shape variation in A-TIG welding process. *Theoretical and Applied Fracture Mechanics*; Vol 48(2), pp 178–186.

Yong Huang, Fan Ding, and Fan Qinghua (2007), Study of mechanism of activating flux increasing weld penetration of AC A-TIG welding for aluminum alloy. *Frontiers of Mechanical Engineering in China*; Vol 2(4), pp 442–447.

Zhao Yong, Yang Gang, Yan Keng, and Liu Wei (2011), Effect on formation of 5083 aluminum alloy of activating flux in FBTIG welding. *Advanced Materials Research*; Vols 311–313, pp 2385–2388.

Index

Printed and bound by CPI Group (UK) Ltd, Croydon, CR0 4YY

23/10/2024

01778243-0001